**W9-CSP-625**

UNDERSTANDING & CONTROLLING AIR POLLUTION

# UNDERSTANDING & CONTROLLING AIR POLLUTION

Howard E. Hesketh, Ph.D., P.E.
Associate Professor of Engineering
Air Pollution Control
Southern Illinois University
Carbondale, Illinois

CHEMISTRY

First Printing, 1972
Second Printing, 1973

Library of Congress Catalog Card Number 72-88892
ISBN 0-250-40007-3

# PREFACE

The use of the word *ecology* has increased rapidly in the past few years--and for good reason. Only a few years ago man found it necessary to exploit the frontier areas so they could become usable farms and cities. The same exploitation must be stopped to regain a proper ecological balance. Every member of the society is affected because all forms of pollution are interrelated. Those of us concerned with solving air pollution problems must not do so by creating water pollution or some other form of pollution. Likewise, other forms of pollution must cease becoming air pollution problems.

Specialists are needed in all aspects of pollution control. *Generalists* who have knowledge of environmental problems are also needed to safeguard against specialist solving of one pollution problem which in turn creates another. Beware of *temporary* improvements that make people believe the problem is solved and needs no further concern, as happened in 1969-70 when air quality in some areas was improved by reductions in massive open burning and control of other extreme emissions. Pollution is *not* under control now, yet it can be and will be, but only as quickly and as efficiently as we want it.

It is important that the specialists and generalists communicate effectively with all segments of the society. A few poets and professional writers (trained in communication) declare that the world is absolutely out of control and it cannot be saved. Scientists (notoriously poor communicators) predict that by the end of this century we can prevent further deterioration and can reverse the process in the direction of melioration. Part of communication is listening, reading, and evaluating to decide which information is correct--the poet's or the scientist's or some combination. If the data being communicated can be checked to ascertain validity before making a decision, the evaluator should assume that responsibility. It is my fervent hope that this volume will assist its user to make these evaluations and thereby aid in solving one aspect of the pollution problem.

Howard E. Hesketh                          September 1972
Carbondale, Illinois

# ACKNOWLEDGEMENTS

Quite frequently the works of the student are merely extensions of ideas received from an influential teacher. The teacher in reference is Dr. Seymour Calvert, who founded the Center for Air Environment Studies at The Pennsylvania State University and who was one of my teachers. I greatly appreciate the influence of this man.

I also acknowledge the assistance provided by the Southern Illinois University which enabled me to prepare this work. Special thanks go to my students who helped proof read and correct the manuscript.

H.E.H.

To Joyce Stern Hesketh

# TABLE OF CONTENTS

## Contents

Contents

Contents

# TABLE OF MOST COMMONLY USED SYMBOLS

*(NOTE: Not all symbols are listed here and some symbols have several meanings; check context as necessary.)*

| | |
|---|---|
| $A^O$ | = Angstroms = $10^{-8}$ cm or $10^{-4}$ μ |
| a | = Acceleration = ft/sec$^2$ or cm/sec$^2$ |
| $C_D$ | = Drag coefficient, dimensionless |
| $C_{(x,y,z)}$ | = Concentration downwind from source at position x, y, z, g/m$^3$ |
| $c$ | = Cunningham correction factor (Eq. 9.7), dimensionless |
| cfm | = cubic feet per minute |
| $D_{AB}$ | = Molar diffusivity of a gas A in gas B, lb moles/(ft hr) |
| $D_{PM}$ | = Diffusivity of particle P through continuous medium M, cm$^2$/sec |
| D | = Diameter |
| d | = Particle diameter, microns |
| d' | = Particle diameter, not in microns |
| $\bar{d}$ | = Arithmetic mean diameter |
| $d$ | = Differential operator |
| $d_{50}$ | = Mean particle diameter which occurs at frequency probability of 50% (Note context to determine whether geometric or arithmetic mean is implied) |
| E | = Overall collection efficiency, % |
| $E_0$ | = Collection efficiency for specified size particle; or electrical field strength |
| exp | = Signifies e (natural log base) to the exponent indicated by the quantity in brackets after exp |

xv

| | | |
|---|---|---|
| esu | = | Electrostatic units |
| g | = | Gravitational acceleration, ft/sec² or dynes/cm² |
| $g_c$ | = | Gravitational acceleration constant, 32.174 ft $lb_m$/($lb_f$ sec²) or 980.7 dynes/cm² |
| H | = | Effective plume height (H' + ΔH), meters |
| H' | = | Stack height, meters |
| ΔH | = | Rise of plume above the stack (positive, negative, or zero), meters |
| h | = | Height of uniformly mixed inversion layer, meters |
| ID | = | Inside diameter |
| $K$ | = | Boltzman constant = $1.38 \times 10^{-16}$ g cm²/ (sec² molecule °K) or may be (per small particle) or $1.38 \times 10^{-23}$ joules/°K) |
| °K | = | Absolute temperature in degrees Kelvin, °C+273.16 |
| K' | = | Pettyjohn shape factor (Eq. 7.6), dimensionless |
| $K_o$ | = | Dielectric constant of a vacuum, $8.8 \times 10^{-12}$ coulombs²/(joule m) |
| L | = | Temperature lapse rate = $-\frac{\partial T}{\partial Z}$ |
| $L_a$ | = | Adiabatic lapse rate = -1°C/1000m or -5.4°F/1000 ft |
| $lb_f$ | = | Pound force |
| $lb_m$ | = | Pound mass (see g and $g_c$) |
| $ln$ | = | Natural logarithm |
| M | = | Molecular weight (also can mean 1,000) |
| $\overline{M}$ | = | Average molecular weight of phase (see Eq. 11.29) |

| | |
|---|---|
| m | = Meter or mass |
| $N_{Re}$ | = Reynolds' number* = $\dfrac{Dv\rho}{\mu}$, dimensionless (see also Re) |
| $N_{Sc}$ | = Schmidt number* = $\dfrac{\mu}{\rho\, D_{PM}}$ or $\dfrac{\mu}{M\, D_{AB}}$, dimensionless |
| $N_c$ | = $\Sigma$ no. of calms |
| $N_e$ | = $\Sigma$ principal wind direction frequencies |
| $N_o$ | = $\Sigma$ secondary wind direction frequencies |
| n | = Sum of numerical values (frequency); or a number |
| $n_e$ | = Frequency for any one particular principal wind direction |
| $n_o$ | = Frequency for any one particular secondary wind direction |
| P | = Total pressure (See Appendix C) |
| $P°$ | = Vapor pressure of pure substance at some given temperature |
| p | = Partial pressure |
| ppm | = Parts by volume per million parts total volume (for ideal gases: vol ratio = mole ratio = pressure ratio) |
| ppb | = Parts per billion |
| psia | = Pounds per square inch absolute = psig + atmospheric pressure in psi |
| psig | = Pounds per square inch gauge |
| Q | = Source strength when pollution is released, g/sec; or volumetric flow rate, $ft^3/min$ |
| R | = Ideal gas law constant (See Appendix C) |
| $°R$ | = Degree Rankine = $°F + 459.49$ |

\* $\rho$ and $\mu$ refer to the fluid phases

Re $\quad$ = Drop Reynolds' number = $\dfrac{d(v_p - v_g)\rho g}{\mu_g}$, dimensionless (see also $N_{Re}$)

$\hbar$ or r $\quad$ = Particle radius or reaction rate

SC $\quad$ = Standard conditions, 60°F and 1 atmosphere unless otherwise noted

STP $\quad$ = Standard temperature and pressure, 32°F and 1 atmosphere

T $\quad$ = Absolute temperature, °R or °K

$\bar{u}$ $\quad$ = Mean wind speed, meters/sec

v $\quad$ = Velocity, ft/sec

$v_a$ $\quad$ = Velocity of air

$v_g$ $\quad$ = Velocity of gas

$v_p$ $\quad$ = Velocity of particle

$v_S$ $\quad$ = Stokes' terminal settling velocity (Eq. 9.6) cm/sec

$X_S$ $\quad$ = Stokes' stopping distance (Eq. 9.11), cm

x $\quad$ = Distance downwind from source, meters; or abscissa of graphs

y $\quad$ = Distance horizontally from plume centerline, meters; or ordinate of graphs

z $\quad$ = Height above ground, meters

## GREEK LETTERS

| | |
|---|---|
| $\alpha$ | = Any number value |
| $\bar{\alpha}$ | = Arithmetic average or mean of |
| $\alpha'$ | = Deviation from the mean |
| $\partial$ | = Partial differential operator |
| $\eta$ | = Effective efficiency, fraction |
| $\mu$ | = Micron = $10^{-3}$ mm; or viscosity |
| $\mu_a$ | = Viscosity of air = $1.8 \times 10^{-4}$ g/(cm sec) = $1.8 \times 10^{-4}$ poise = $1.21 \times 10^{-5}$ lbm/(sec ft) or = $3.76 \times 10^{-7}$ lb$_f$sec/ ft$^2$ at SC |
| $\mu_g$ | = Viscosity of gas |
| $\pi$ | = 3.1416 |
| $\rho$ | = Density, lb/ft$^3$ or g/cm$^3$ |
| $\rho_a$ | = Density of air = $1.2 \times 10^{-3}$ g/cm$^3$ = $7.50 \times 10^{-2}$ lb/ft$^3$ at SC |
| $\rho_g$ | = Density of gas |
| $\Sigma$ | = Summation |
| $\sigma$ | = Standard deviation |
| $\sigma_y$ | = Horizontal (cross wind) deviation, meters |
| $\sigma_z$ | = Vertical deviation, meters |
| $T$ | = Surface tension = $(0.04)(641 - °K)^{1.28}$ dyne/cm for water near normal SC |
| $T_{zu}$ | = Downward shear stress momentum in down wind direction acting on the wind in the z plane |

# PART I
## GENERAL CONSIDERATIONS

# CHAPTER I

# AIR POLLUTION AND SOCIETY

Air pollution can be controlled but society must decide to what level and when it should be controlled. In the first half of this book, it is hoped that the reader will gain an understanding about what air pollution is and what it does. For those concerned with how air pollution can be controlled, Part II of this book will be very useful. The purpose of Chapter I is to orient the reader so that he can understand what efforts are being undertaken and perhaps more importantly, it may help point out what is not being done.

## 1.1 AWARENESS

Odors are the most common source of air pollution complaints. Most of what we call air pollution could be roughly classified as either smoke or odors.

As early as 1300, a royal decree was issued in London prohibiting the use of low-grade coal for heating because it created excessive smoke and soot. The only known case of capital punishment because of an air pollution violation occurred in the 13th Century when a Londoner violated this order. Sulfur in fuels burns to sulfur dioxide. In 1600 sulfur dioxide was the first chemical to be specifically recognized as an air pollutant. However, it was not until about 1940 when air pollution, as such, became important.

It should be noted that the earth, which had been warming up prior to 1940, is now cooling. Artic winter temperatures have in fact dropped an average of 6°F. It is theorized that this is due to the air pollution in the atmosphere, in particular to the more than 10% increase in the carbon dioxide content since 1900. This cooling effect may even be further intensified due to

the changing of the reflectivity of the earth which
is being altered by jet contrails.  A Boeing 707
burns one ton of fuel every ten minutes, releasing
1.3 tons of water vapor and 3.2 tons of $CO_2$ plus
other gases.  Most of the $CO_2$ is released in the jet
airstreams and years are required for this material
to enter the lower elevation atmospheric circulation
system and thereby become incorporated back into
the biological carbon cycle.  It is not apparent
what effects this continuing activity will have on
the earth's life.  Visible portion of the jet
contrails are condensed water droplets that form
ice at high altitudes.  This produces 30 to 40 days
per year of cirrus cloud cover in areas where there
are jet airplane lanes.

Automobile exhaust contributes over 60% of the
total air pollution that now exists in the atmosphere.
In addition to the direct results of the pollution,
it is also possible that particles released with the
exhaust, such as lead oxide, super-seed the clouds
making them unable to release their rain.  This
could be an explanation for the high number of
droughts which have occurred in various portions of
the United States.

Authors of technical papers presented at the
1969 Air Pollution Control Association meeting wrote
independent responses which uniformly concluded that
increased public awareness of the pollution problem
is the single biggest factor in helping to foster
solutions to excessive pollution (1).  Should the
reader desire supplementary basic air pollution
information he is referred to the "Air Pollution
Primer" (2).

## 1.2   AIR POLLUTION DEFINED

Air pollution is the presence of foreign matter
(either gaseous or particulate or combinations of
both) in the air which is detrimental to the health
and/or welfare of man.  This definition enables us
to include not only the direct effects of air
pollution on man, but the effects of air pollution
which damages materials and reduces the esthetic
value of antimate and inanimate matter.  Remember
that health can also be damaged by mental attitude
and this attitude is affected by factors such as
esthetic and monetary considerations.

Air pollution would not exist if it were not
for the chain which consists of source-transport-
receptor.  If any one of these links were missing,
we would not be affected by pollution.

## 1.2.1  PARTICULATES

In the definition or air pollution, we included
the word matter.  Classed as matter are both particulate and gaseous substances.  Particulates consist
of solid and/or liquid particles.  Particulates
larger than 50 microns settle out of the air quite
easily, and cause two major problems:  (a) deposition,
and (b) adhesion.  Deposition refers to dirt on
clothing, homes, and other property, while adhesion
refers more specifically to respiratory tissue
adhesion.  This will of course cause both physiological, as well as psychological damage.  Most of the
damage due to adhesion in the respiratory tract is
caused by particles from 0.5 to 2.0 microns (μ) in
diameter.  Unfortunately, these size particles are
also the most difficult particles to remove from
a gas stream by pollution control equipment.
Larger particles are removed by inertial
impaction mechanisms and particles smaller than
this are removed by diffusion processes.  Figure
1.1 is a pictorial representation of this showing
that collection efficiency is lowest for approximately 0.5μ particles.

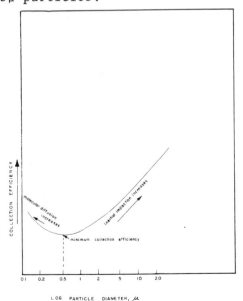

Figure 1.1  General Collection Efficiency by Known
Mechanisms v/s Particle Diameter

## 1.2.2  AEROSOLS

Aerosols are a special class of particulate. They consist of colloidal suspension that are larger than molecular size yet not large enough to settle under gravity. They are often considered to be from 0.01 to 50 microns in size. Particles larger than 50μ in diameter quickly settle out of the atmosphere because of gravity. Aerosols tend to remain suspended for relatively long periods of time. Examples are dust, mist, smoke, fog, haze, and fumes. Aerosols are formed by any of the following three methods:  1) emitted as an aerosol,  2) evolved from breakup of larger particulates (dispersion aerosols) and 3) formed by condensation of nucleation (called fumes if approximately sub-micron in size).  In an aqueous sol type colloidal solution, all particles have different charges:  positive, negative, or neutral.  As implied earlier, it is aerosols that are breathed in and deposited in the respiratory tract.  These aerosols may also contain an electrical charge which can be beneficial or detrimental to man's health (as described in Section 5.2.5).

Aerosols react to both heat and light.  They are repelled by heat and move toward light while others move away.  Another important characteristic of aerosols is the light-scattering effects.  The greenhouse effect is attributed to the presence of aerosols in the atmosphere.  The energy of the sun, which is in wave lengths between 0.1 and 30 microns, travels through the earth's atmosphere and is absorbed by the atmosphere with its pollutants and by the surface of the earth.  The earth then re-radiates energy in accordance with Plank's Law (proportional to the fourth power of the earth's temperature), giving off energy in about a 10 micron wave length.  More of the re-radiated energy is absorbed by the atmosphere resulting in a net conservation of heat energy.  This buildup is known as the greenhouse effect.  As previously mentioned, the earth's temperature has been cooling down since 1940, so we do have the offset of the greenhouse effect due to the "clouding over" effect of the atmospheric pollutants.

The following is a summary of aerosol effects based on adverse atmospheric and physiological effects for various approximate size ranges:

0.1μ - max. cloud nucleation phenomena and possible weather modification

0.4 - 0.8μ - max. light scattering and visibility restriction (these diameters equal the wave length of visible light)

0.5 - 2μ - max. deposition in the lungs

10μ - result in most of the dustfall deposits of dirt

## 1.2.3 GASES

We know that gaseous matter occupies space and has weight and that gases expand by diffusion until the molecules are equally distributed within a localized area. Air pollutants are usually present at very low concentrations so we do not distinguish between gases and vapor; however, vapors are closer to the boiling point curve and deviate more from ideal gas laws. Air behaves as an ideal gas at ambient conditions and the fact that the concentration of gaseous pollutants in the atmosphere is low makes it possible to consider them as ideal gases also. Table 1.1 shows the composition of clean, *dry* air near sea level, in both percent by volume and parts per million. Note that polluted urban air can vary substantially from these values. For example, methane and carbon monoxide in urban air usually have a concentration of from 6 to 9 ppm.

Gases 1) cause odors, 2) react chemically with other pollutants in the atmosphere to form secondary pollutants which can be either (or both) gases or aerosols, 3) chemically irritate living tissue, and 4) react with and destroy inanimate objects. Some gases cause only one of these problems while others cause many. The extent of the damage varies with the concentration of the particular pollutant in the atmosphere and natural factors (health, sunlight, etc.).

One of the most common pollutants, smoke, is classified as an aerosol, yet it contains gaseous pollutants, as well. The aerosol particles of smoke are in the size range from about 0.03 to 10 microns in diameter.

TABLE 1.1   TYPICAL COMPOSITION OF CLEAN, DRY AIR
NEAR SEA LEVEL

| Component | Formula | Content % by Vol. | Content ppm |
|----------|---------|-------------------|-------------|
| Nitrogen | $N_2$ | 78.09 | 780,900 |
| Oxygen | $O_2$ | 20.94 | 209,400 |
| Argon | Ar | 0.93 | 9,300 |
| Carbon dioxide | $CO_2$ | 0.033 | 330 |
| Neon | Ne | $18 \times 10^{-4}$ | 18 |
| Helium | He | $5.2 \times 10^{-4}$ | 5.2 |
| Methane | $CH_4$ | $1.5 \times 10^{-4}$ | 1.5 |
| Krypton | Kr | $1.0 \times 10^{-4}$ | 1.0 |
| Hydrogen | $H_2$ | $0.5 \times 10^{-4}$ | 0.5 |
| Nitrous oxide | $N_2O$ | $0.5 \times 10^{-4}$ | 0.5 |
| Xenon | Xe | $0.08 \times 10^{-4}$ | 0.08 |
| Ozone* | $O_3$ | $0.07 \times 10^{-4}$ | 0.07 |
| Ammonia | $NH_3$ | $0.01 \times 10^{-4}$ | 0.01 |
| Iodine | $I_2$ | $0.01 \times 10^{-4}$ | 0.01 |
| Nitrogen dioxide | $NO_2$ | $0.001 \times 10^{-4}$ | 0.001 |
| Sulfur dioxide | $SO_2$ | $0.0002 \times 10^{-4}$ | 0.0002 |
| Carbon monoxide | CO | 0 to trace | — |

*Ozone content in winter is 0.02 ppm and in summer is 0.07.

1.3  AIR QUALITY

Air quality varies from one location to another
depending on:  1) geographical location, 2) meteorological conditions, 3) industrial proximity,
4) pollution type, and 5) size of the community (as
shown in Section 1.2 by the urban concentration of
methane).  An example of geographical variations is
that certain midcentral United States areas have
high non-urban particulate concentration--over 40
micrograms per cubic meter while the northcentral
portion of the United States may have particulate
concentrations of less than 10 micrograms per cubic
meter.  These are natural, non-manmade sources of
pollution.  To these natural sources we must add the
manmade sources, the sum of which then becomes the
total particulate concentration in the ambient air.
Table 1.2 shows average *urban* particulate concentration as summarized from data of the National Air
Sampling Networks which consists of up to 127 urban
sampling areas.  Other geographical factors such as
mountain valley regions result in periods of poor
air quality and are discussed in Chapter III.

Most manmade pollutants have a short life in the
atmosphere and remain there for only a few hours or
at most for a few weeks.  Carbon monoxide and some
organic compounds such as methane differ from this
normal situation and have life times of several years.
(It is estimated that over $2.2 \times 10^9$ tons of
methane are released into the atmosphere annually
from Middle East oil wells.)  Carbon monoxide and

TABLE 1.2  AVERAGE U.S. URBAN PARTICULATE
           CONCENTRATION
           *micrograms per cubic meter*

| Year | Geometric Mean | Arithmetic Mean |
|------|----------------|-----------------|
| 1957 | 115 | 134 |
| 1958 | 108 | 127 |
| 1959 | 101 | 122 |
| 1960 | 99 | 115 |
| 1961 | 92 | 106 |
| 1962 | 92 | 108 |
| 1963 | 92 | 108 |
| 1964 | 94 | 109 |
| 1965 | 87 | 102 |
| 1966 | 92 | 103 |
| 1967 | 89 | 101 |

water vapor released from high flying aircraft may take as long as 25 years before natural circulation can return them to the lower level of the atmosphere and they enter into the biological life cycle.

Meteorological conditions such as wind, rain, amount of sun, and other factors cause variations in air quality. The amount of air pollution released as well as the proximity of air pollution contributors in areas which have poor dilution capabilities due to meteorological conditions will result in very poor ambient air quality.

Quality of the local air is influenced by the proximity of large size utilities and industries, but it also is affected by sources such as vehicle density, residential heating, open incineration, natural releases (swamps, mines, volcanoes), etc. For example, California has more vehicles per square mile than any other area of the United States. This results in a concentration of nitrogen oxides from vehicle exhaust which is higher in this area than in any other area. Also, the northcentral and northeast portions of the United States burn more fuel for home heating than any other area of the country. The sulfur burned in the fuels helps create the high sulfur dioxide concentrations that exist in this area.

It is important to remember that the ambient air quality depends on both natural and manmade emissions and even if all the manmade contributions were removed it could still be unhealthful. In addition to affecting health, bad air can damage vegetation, affect inanimate objects, and reduce visibility. All of this can occur at extremely low pollutant concentrations. The total volume of contaminants in polluted air, including both gaseous and particulate, is often less than 0.1 percent.

## 1.4  LEGAL ASPECTS

Most air pollution control activities began with the Clean Air Act of 1963. This act was amended in 1965, 1966, 1967, and 1970. The 1967 amendments, known as the Federal Air Quality Act, were themselves amended in 1968 to provide a basis for federal and state governments to develop air pollution control programs on a regional scheme and to establish the issuing of criteria and control documents. The federal government (Department of Health, Education and Welfare) then began designating the geographical boundaries of local air quality regions across the nation. These regions are both interstate and intrastate so that consistent air quality can be obtained in a designated area even though this area includes several states.

The Department of Health, Education and Welfare, as a result of the 1967 Clean Air Amendment, began publishing air quality criteria and control techniques. The criteria summarize all currently available data concerning a given pollutant or group of pollutants. The criteria also list the adverse effects of the pollutants for various concentrations and exposure times. The criteria are documented summaries of available data and attempt to provide the most complete information possible on specific pollutants. The published control techniques describe various methods of controlling the pollutants.

Even though the criteria contain as much data as it is possible to collect, there is still need to interpret and extrapolate the data. The first air quality criteria published were those for sulfur dioxide and particulates, on February 11, 1969. Criteria for carbon monoxide, photochemical oxidants, and hydrocarbons were published in 1970 and criteria for nitrogen oxides were published in 1971. It is expected that other criteria will be issued for substances such as lead, fluorides, polynuclear organic matter, odors and hazardous metals.

Independent of the Clean Air Act, the federal Environmental Protection Agency (EPA) was established on December 2, 1970 (by the same act that formed the National Oceanic and Atmospheric Administration). The EPA supervises the activities of air pollution control, water pollution control, solid waste disposal, radiation safety functions, pesticide control and ecological research. *Formerly*, these pollution control activities had been set up as follows: air pollution control activities were supervised by the Department of HEW through the National Air Pollution Control Administration (NAPCA); solid waste disposal, water hygiene control and radiation safety control were under the supervision of the Department of HEW: water pollution control was under the supervision of the Department of Interior; pesticide programs were under the super-vision of the Food and Drug Administration and the Departments of Agriculture and Interior; radiation protection was under the supervision of the Atomic Energy Commission; and ecological research was a function of the Council on Environmental Quality. The Office of Air Programs (OAP) of EPA is

essentially the former NAPCA.  The OAP-EPA budget
request for the 1973 fiscal year (starting July 1,
1972) is for 119.7 million dollars which amounts
to about 4.8% of the total EPA requested budget.
The funds are requested for air pollution work in
1973 as follows (in millions of dollars):

    Research, Development, and Demonstration
        Control technology - - - - - - - 30.6
        Regional studies - - - - - - - - 5.0
        Health effects - - - - - - - - - 11.8

    Abatement and Control
        Assistance to control agencies - - 51.5
        Compliance and surveillance - - - 8.5

    Enforcement
        Enforcing standards - - - - - - 2.3

    The EPA has established ten Federal Regions,
each headed by a regional director to coordinate
and supervise the environmental activities in each
area.  Table 1.3 lists these Federal Regions.
    The Clean Air Amendments of 1970 (Public Law
91-604) were signed on December 31, 1970.  This
complicated act is said to have more deadlines per

TABLE 1.3  ENVIRONMENTAL PROTECTION AGENCY FEDERAL
           REGIONS

| Region | Includes these states |
|--------|----------------------|
| I | Maine, New Hampshire, Massachusetts, Vermont |
| II | New York, New Jersey |
| III | Pennsylvania, West Virginia, Virginia, Delaware, Maryland, District of Columbia |
| IV | Kentucky, Tennessee, N. Carolina, Mississippi, Alabama, Georgia, S. Carolina, Florida, Virgin Islands |
| V | Minnesota, Wisconsin, Illinois, Michigan, Indiana, Ohio |
| VI | New Mexico, Texas, Oklahoma, Arkansas, Louisiana |
| VII | Nebraska, Kansas, Iowa, Missouri |
| VIII | Montana, N. Dakota, Wyoming, S. Dakota, Utah, Colorado |
| IX | California, Nevada, Arizona, Hawaii, Pacific Territories |
| X | Washington, Oregon, Idaho, Alaska |

square inch than any bill passed in the last 20
years. It provides deadlines for the federal and
state governments and for industry. The most
significant points of this act are the planning for
national ambient air quality standards, setting
automobile emissions limits, scheduling implementation deadlines and providing funds for research.
These points are explained in Section 1.5.

Air quality standards are not enforceable
(i.e., the air cannot be told to be clean). Only
emission limiting standards can be enforced. Proper
specification of emission standards to assure
attaining the air quality standards may not be easy,
and trial and error methods may be the quickest
procedure.

In addition to the requirements of the Clean
Air Act, most states and many counties and cities
have enacted legislation for pollution control.
In general, any lower form of government may
establish air pollution control legislation as long
as it is as stringent as, or more stringent than
legislation enacted by the next higher governmental
agency. A summary of basic state and local laws is
presented annually by "Chemical Engineering". It
lists by states the current air pollution control
legislation.

## 1.5  CURRENT ACTION

Public Law 91-604 requires the EPA administrator
to establish Federal Air Quality Standards. Standards for six pollutants were issued in April, 1971--
others will be published later. The states must
meet these standards before the end of 1975. The
first Federal Air Quality Standards are:

### $SO_2$
 80 $\mu g/m^3$ (0.03ppm) annual arithmetic mean
365 $\mu g/m^3$ (0.14ppm) 24 hr. max. once a year

### Particulate
 75 $\mu g/m^3$ annual geometric mean
260 $\mu g/m^3$ 24 hr. max. once a year

### CO
10 $mg/m^3$ (9ppm) 8 hr. max. once a year
40 $mg/m^3$ (35ppm) 1 hr. max. once a year

### Photochemical Oxidants
160 $\mu g/m^3$ (0.08ppm as $O_3$) 1 hr. max. once a year

100 $\mu g/m^3$ (0.05ppm as $\frac{NO_x}{NO_2}$   annual arithmetic mean

160 $\mu g/m^3$ $\frac{Hydrocarbons\ (non\text{-}methane)}{(0.24ppm\ as\ CH_4)\ 3\ hr.\ max.}$ once a year
(6-9 AM)

   The Clean Air Amendments of 1970 require the
EPA administrator to determine emission levels for
certain dangerous substances such as cadmium,
beryllium, mercury, asbestos, arsenic, chlorine,
hydrogen chloride, copper, manganese, nickel, zinc,
barium, boron, chromium, pesticides, radioactive
substances and others.
   The states must submit implementation plans to
the EPA for approval.  These plans are to include
the emission requirements for all pollution sources
so that the ambient air quality standards can be
achieved.  If the EPA rejects the plans, it must
provide substitute plans for the state.
   The 1970 law requires that automobile manufacturers provide 5 years or 50,000 miles warranties
on pollution control devices by 1972.  The industries
also have until 1975 (or 1976 if the 1 year extension
is made) to produce automobiles with emission levels
90% less than the 1970 automobile emissions.
   Penalties for knowingly violating this law are
up to $25,000 per day and/or up to 1 year in jail
for first offenses and double for second violations.
In addition, if the EPA administrator fails to
discharge his legal responsibilities, the federal
government can be sued under this act.
   The annual budget for the U.S. is currently
over $200 billion.  Air pollution control activities
receive 0.00005% of these monies.  At least, the
new law will improve this substantially by providing
1.1 billion dollars for research and enforcement
for the period 1971-1973.  In addition, some states
and the federal government provide financial
assistance and/or tax relief (rapid write-off) for
pollution control.  In the contiguous states,
currently 37 states provide financial assistance
and 29 states provide some form of tax relief.
Information concerning details for individual
states can be obtained from the capitol of the
state concerned, the IRS, and the Federal Control
Region Coordinator.

QUESTIONS FOR DISCUSSION

1. Compare the definition for "air pollution" with the federal and state definitions.
2. Repeat #1 for particulate matter. Is it important to consider condensable material here?
3. What are the state emission standards?
4. How much money or approximate percent of the United States budget is currently being spent on air pollution control by the federal government and through what agency is it distributed?
5. What type of tax relief is provided on pollution control equipment by the state?

REFERENCES

1. Hesketh, H.E. "Air Pollution - A Moral Issue", All Clear - The Magazine for Environmental Administration, Vol. 2, No. 3, pp. 19-20 (1970).
2. "Air Pollution Primer", National Tuberculosis and Respiratory Disease Association, New York, N.Y., 103 pp. (1969).

# CHAPTER II

## SOURCES AND EMISSIONS

Pollutants will be discussed in this chapter according to the five source headings: transportation, industry, power generation, space heating, and refuse burning. Approximately 90% by weight of this pollution is gaseous and 10% is particulate matter. In 1968, it was estimated that over 214 million tons of pollutants were released into the U.S. atmosphere.

Pollutants could be further classified according to their physical and chemical composition as shown by the following examples:

inorganic gases - sulfur dioxide, hydrogen sulfide, nitrogen oxides, hydrochloric acid, silicon tetrafluoride, hydrogen fluoride, ammonia, ozone,...

organic gases - terpines, mercaptans, hydrocarbons,...

inorganic particulates - lime, metal oxides, silica,...

organic particulates - pollen, smuts, fly ash,...

## 2.1  TRANSPORTATION

Transportation source pollutants released into the atmosphere account for over 60% of the total U.S. air pollution. In 1970, about 100 million tons of pollutants were released by transportation sources. The pollution composition by weight is 82% carbon monoxide (CO), 11% hydrocarbons (HC) (this represents 4.7% by weight of the hydrocarbon fuel used (1)), 4% nitrogen oxides ($NO_x$), 2% particulate matter, and 1% sulfur dioxide ($SO_2$). This comes from the approximately 100 million motor vehicles in the United States. Most of the internal combustion engines are 4 cycle automobile engines, but also included are motorcycles,

TABLE 2.1  PARTIAL LISTING OF INDUSTRIAL SOURCES OF AIR POLLUTION AND SOME OF THEIR EMISSIONS, 1971

| Source | U.S. Production in 1000 tons, Est. | Some Emissions | Typical Uncontrolled Factors lb/ton prod. |
|---|---|---|---|
| **CHEMICALS** | | | |
| Acids (100% basis) | | | |
| Hydrochloric | 2,025 | HCl | 3 |
| | | $Cl_2$ | 2 |
| | | Organic vapors | 15 |
| | | Catalyst dust | 0.1 |
| Hydrofluoric | 325 | HF, F | 20 |
| | | $SO_x$ | 0.05 |
| | | Fluorspar dust | 25 |
| Nitric | 6,671 | $HNO_3$ mist | 11 |
| | | $NO_2$ | 40 |
| | | NO | 8 |
| | | $NH_3$ | 1 |
| Phosphoric ($P_2O_5$) | 6,034 | $F^-$, HF | 23 |
| | | $SO_2$ | 2 |
| | | Gypsum dust | 10 |
| Sulfuric | 29,285 | $H_2SO_4$ mist | 5 |
| | | $SO_2$ | 40 |
| | | $SO_3$ | 5 |
| | | S | 5 |
| Alkalies | | | |
| Potassium Hydroxide (90%) | 187 | KOH | 20 |
| | | $Cl_2$ | 35 |
| Sodium Hydroxide (58% $Na_2O$) | 9,683 | (Produced with chlorine) | (See chlorine) |

| Product | Production | Pollutant | Amount |
|---|---|---|---|
| Chlorine | 9,341 | $Cl_2$ | 50 |
| | | $H_2$ | 5 |
| | | CO | 50 |
| | | Hg vapor | 0.01 |
| | | NaOH mist | 10 |
| **Inorganic Fertilizer Materials** | | | |
| Ammonia | 13,719 | $NH_3$ | 150 |
| | | $H_2$ | 10 |
| | | $CH_4$ | 100 |
| | | CO | 100 |
| Ammonium Nitrate | 6,584 | $NH_4NO_3$ | 60 |
| | | $NH_3$ | 50 |
| | | $HNO_3$ | 5 |
| Ammonium Sulfate | 2,325 | $(NH_4)_2 SO_4$ | 50 |
| | | $NH_3$ | 50 |
| | | $H_2SO_4$ | 5 |
| | | $SO_2$ | 1 |
| **Rubber** | | | |
| Synthetic | 2,079 | Organic vapors | 55 |
| | | Particulates | 10 |
| Natural | 602 | Organic vapors | 60 |
| | | Particulates | 25 |
| **Syndets** | | | |
| Soap | 470 | Organic vapors | 20 |
| | | Particulates | 25 |
| Synthetic Detergents | 2,600 | Particulates | 60 |
| | | Organic vapors | 5 |

TABLE 2.1  CONT.

| Source | U.S. Production in 1000 tons, Est. | Some Emissions | Typical Uncontrolled Factors lb/ton prod. |
|---|---|---|---|
| PETROLEUM AND COAL | | | |
| Petroleum (Crude Oil) | | | |
| Catalytic Cracking Units | 548,000 | Particulates | 0.5 |
| | variable | $SO_x$ | 2 |
| | | CO | 50 |
| | | Organic vapors | 1 |
| | | $NO_x$ | 0.5 |
| | | $NH_3$ | 0.2 |
| | | $H_2S$ | 0.001 |
| | 109 | Particulates | 200 |
| | | Organic Vapors | 100 |
| | | $SO_2$ | 25 |
| | | CO | 100 |
| | | $NH_3$ | 10 |
| | | $H_2S$ | 5 |
| METALS | | | |
| Ore roasting | variable | Metallic dusts | 50 |
| | | Gases (e.g. F or $SO_x$) | 100 |
| Ferroalloy Electric Arc Furnace | | Particulates | 300 |
| | | CO | 25 |
| Iron Blast Furnaces | 132,000 | Particulates | 150 |
| | | CO | 30 |

| Source | Emissions | Pollutant | Value |
|---|---|---|---|
| Steel Furnaces (Electric Arc, Open Hearth, Basic Oxygen) | 125,000 | Particulates | 10–50 |
| | | CO | 3–20 |
| Non-Ferrous Furnaces | | | |
| Aluminum | 3,930 | | |
| Copper | 2,220 | Particulates | 10–120 |
| Lead | 1,390 | $SO_x$ | 750 |
| Zinc | 740 | | |
| Magnesium | 124 | | |
| Re-Melting Furnaces | | | |
| Grey Iron Cupolas | variable | Particulates | 15 |
| | | CO | 150 |
| Secondary Lead Smelting | variable | Particulates | 100 |
| | | $SO_x$ | 90 |
| Secondary Zinc Smelting | variable | Particulates | 45 |
| Plating | variable | Acid & Metal Fumes | 2 |
| MINERALS | | | |
| Asphalt Concrete | 25,000 | Particulates | 30 |
| | | Organic vapors | 5 |
| Cement Plants | 79,000 | Lime, limestone & sulfate dusts | 250 |
| Ceramic and Brick Manufacture | | Particulates | 180 |
| Crushing and Classifying Rock | variable | Particulates | 45 |
| Grinding and Polishing | variable | Abrasives & metal dusts | 10 |
| Gypsum Plants | | Particulates | 100 |
| | | $SO_x$ | 5 |

TABLE 2.1 CONT.

| Source | U.S. Production in 1000 tons, Est. | Some Emissions | Typical Uncontrolled Factors lb/ton prod. |
|---|---|---|---|
| MISCELLANEOUS | | | |
| Farming | | | |
| Pesticides | 500 | Particulates | 60 (1) |
| Fertilizers | 18,460 | Particulates | 25 |
| Crushing, Screening and Shelling | 80,000 | Particulates | 1,000 (3) |
| Soil Tilling | 200 (2) | Particulates | variable |
| Natural Releases (e.g. Pine Terpines and Pollens) | variable | Organic vapors | variable |
| Plant and Animal Products | | | |
| Wood Working | 94,500 | Particulates | 5 |
| Grain Processing | 120,000 | Particulates | 10 |
| Sugar Refineries | 10,600 | Particulates | 4 |
| Kraft Wood Pulping | 32,000 | Particulates | 200 |
| | | $SO_x$ | .5 |
| | | CO | 70 |
| | | $H_2S$ | 10 |
| | | Mercaptans | 1 |
| Surface Finishing | | | |
| Paint, Varnish, and Lacquer | 431 | Organic vapors | 100 (4) |
| | | Particulates | 2 (4) |
| Industrial Finishes | 443 | Organic vapors | 100 (4) |
| | | Particulates | 2 (4) |

Waste Disposal
Incineration

|  | | |
|---|---|---|
| Up to<br>1800 lb/person | Particulates | 1-70 (1) |
| | $SO_x$ | 1 (1) |
| | CO | 1-200 (1) |
| | $NO_x$ | 1-10 (1) |
| | Organic vapors | 1-40 (1) |

Decay (Aerobic and Anerobic)

| variable | Particulates | variable |
|---|---|---|
| | CO | variable |
| | Organic vapors | variable |

(1) lb/ton used (i.e. lb/ton process weight)
(2) millions of acres tilled
(3) lb/acre year
(4) lb/ton solvent

lawnmowers, outboard motors and others (of which
some are two cycle as well as four cycle). Compression ignition engines (diesel type engines), turbines,
and jets are also internal combustion engines and are
discussed in Chapter VI.

## 2.2 INDUSTRY

In 1970, about 26 million tons of pollutants
were released by industrial sources. This mass of
pollution represents from 0.25% to 3% of the
industrial material being processed at a specific
plant location (1). The composite by weight of
industrial emissions is approximately 33% $SO_2$, 26%
particulate matter, 16% HC, 11% CO, 8% $NO_x$ and 6%
other emissions. The industrial sources, by far,
represent the most diversified type of pollution
emissions. Certain of the pollutants are characterized
by a specific color which is associated with the
particular chemical emitted. For example, during
open-hearth furnace steel making, orange fumes of
iron oxide ($Fe_2O_3$) are emitted. Nitric acid manufacturing operations frequently emit a reddish-brown
colored plume of nitrogen dioxide ($NO_2$). This
becomes more brownish, then yellowish and finally
colorless as it changes to nitric oxide (NO) and
oxygen.
Table 2.1 is a partial listing of some typical
industrial operations showing some of the pollutants
released and the amount that could be released if
the operation had no pollution controls. Exact
amounts and types of emissions will vary depending
on factors including specific process method and
facilities, operating techniques, raw material
characteristics (physical size and shape and/or
chemical purity), product grade and weather and
climatic conditions. Table 2.1 is an attempt to
show average uncontrolled process emissions and in
most cases includes typical losses due to handling
of raw materials and products. These data were
assembled using emission data available from the EPA
and personal records plus extrapolations of production data from the Department of Commerce, Tariff
Commission and Bureau of Mines using the Federal
Reserve System Industrial Production Indices.
Industrial operations are continually changing
so the types and amounts of emissions also change.
An example is the steel industry where the open-
hearth furnace is being phased out and the use of
the basic oxygen furnace (BOF) is increasing. The

oxygen furnace has the advantage of reducing the
time of the heat from approximately three hours to
thirty minutes. These enclosed furnaces confine
the gaseous emissions making it easier to control
the emissions. One of the disadvantages of this
operation is the fact that the BOF uses only about
28% scrap per charge compared with the 50% scrap
that could be used in the old open-hearth charge.

Table 2.2 shows how much money may have to be
spent by various industries to meet the provisions
of the HEW criteria as outlined in the Clean Air
Act. By the year 1975, the costs to reduce
emissions of particulates, sulfur oxides, hydrocarbons and carbon monoxide for the manufacturing
industries listed will range from a low of $5 million
per year for phosphate fertilizer plants and
petroleum refineries to a high of $823 million per
year for the iron and steel plants. On the average,
all plants in the United States in 1970 spent about
0.8% of their annual income on air pollution control.

## 2.3  POWER GENERATION

Approximately 22 million tons of pollutants
were released into the atmosphere in 1970 from
generation of electric and steam power. This
pollution represents 0.05% to 1.5% by weight of
the fuel consumed (1). The composition of this
pollution by weight is approximately 58% $SO_2$, 5% CO,
5% HC, 17% particulate matter and 15% $NO_x$. Most
power generation plants burn fossil fuels. (Fossil
fuels are those fuels which come from fossilized
organic matter and are mostly coal, natural gas and
oil.) Fossil fuels now produce 85% of the nation's
electricity but they also contribute 50% of the
sulfur oxide emissions, 25% of the particulate and
25% of the nitrogen oxide releases. Combustion of
fossil fuels by all processes has been largely
responsible for the increase in the atmospheric $CO_2$
concentration. During the past 50 years, $CO_2$
concentration has increased from 0.030 to 0.033%,
a 10% increase. This is spectacular when one
considers that the world wide annual $CO_2$ emission
rate amounts to about $1.3 \times 10^{10}$ tons from manmade
operations and $1 \times 10^{12}$ tons from natural sources.
Current National Bureau of Standards reports
conclude that the oxygen content in the atmosphere
is the same as it was 60 years ago (20.946%) and
that the carbon monoxide concentration is not
increasing although approximately $200 \times 10^6$ tons/yr.

TABLE 2.2   INDUSTRIAL AIR POLLUTION CONTROL COSTS AS PROJECTED BY HEW AND MCGRAW-HILL DEPARTMENT OF ECONOMICS (based on 1967 plants)

| Industrial Plant | 1975 Emissions, M tons/yr | Required Emission Reduction | Control costs, in $10^6$ Capital Investment | Total Annual Costs |
|---|---|---|---|---|
| Phosphate fertilizer | 16 particulates | 67.0% | 5 | 5 |
| Petroleum refineries | 63 " | 45.5% | | |
| | 1130 $SO_2$ | 20.6% | 103 | 5 |
| | 719 HC | 86.4% | | |
| | 1323 CO | 91.0% | | |
| Kraft pulp (sulfate) | 164 particulates | 80.5% | 7 | 7 |
| Sulfuric acid | 55 " | 40.0% | | |
| | 561 $SO_2$ | 86.0% | 37 | 7 |
| Rubber (tire & related) | 65 particulates | 90.2% | 55 | 59 |
| Iron & Steel | 1420 " | 90.5% | 412 | 823 |
| Chemicals | | | 223 | |
| Paper | | | 140 | |
| Rubber | | | 64 | |
| Petroleum | | | 208 | |
| Food & Beverages | | | 69 | |
| Textiles | | | 2 | |
| All manufacturing | | | 1,901 | |

of carbon monoxide are released from all sources.
More  accurate analytical measurements are expected
to show significant increases in CO levels.  Supplementary and alternate fuels for power generation
include: hydroelectric systems, fuel cells, atomic
energy and natural sources (tide, wind, solar and
geothermal energy and magnetic force).

The fossil fuels, which are used to produce
the approximately 1500 billion kilowatt hours of
electricity in 1970, contain sulfur as the main
impurity.  Other impurities include heavy metals
such as mercury, cadmium, and beryllium.  A sulfur
dioxide plume from fuel burning appears as a bluish-
white plume and should not be confused with the
brilliant white plumes of condensed water vapor
which dissipate rapidly.

Figure 2.1 shows that since 1940, the use of
oil has varied while the use of coal and natural
gas have greatly increased.  The sources for electric
power generation in the U.S. in 1969 are fossil
fuels 82.4%, hydropower 16.7% and nuclear power
0.9% (2).  Coal used by electric facilities amounted
to $310.6 \times 10^6$ tons in 1969 and is estimated to be
$330 \times 10^6$ tons in 1970 (bituminous coal accounts
for 97% of the total coal used for electric power
generation).  Electric utilities used 58% of the
total bituminous coal consumed in 1969 with industry
using another 37%.  Natural gas production increased
7% from 1968 to 1969 with $3.5 \times 10^{12} ft^3$ natural gas
going to electric utilities.  The 1969 consumption
on natural gas was divided as follows:  19% to
electric utilities; 25% to residential; 10% to
commercial; and 46% to industrial.

Figure 2.2 shows that most of the $SO_2$ comes
from coal, but combustion of petroleum products
also releases $SO_2$.  Figure 2.3 shows the increase
in nitrogen oxide production.  Coal produces more
$NO_x$ per BTU than oil and gas produces less.  Hydrocarbon pollution from power generation comes mainly
from natural gas combustion as shown in Figure 2.4.
Coal releases very little hydrocarbon when burned
(0.3 lb/ton) but the large amount of coal burned
results in the emissions shown.  The values
presented in Figures 2.2 - 2.4 were calculated
using data from (2) and (3) and assuming 3.5% sulfur
coal and 2% sulfur oil.

The total energy sources for the U.S. in 1969
can be listed as:  petroleum 43.2%; natural gas
32.1%; bituminous coal (and lignite) 20.1%; water
4.0%; anthracite coal 0.4%; and nuclear power 0.2%.

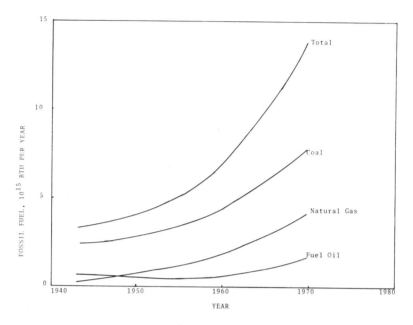

Figure 2.1   Fossil Fuel Used in Electric Power
             Generation

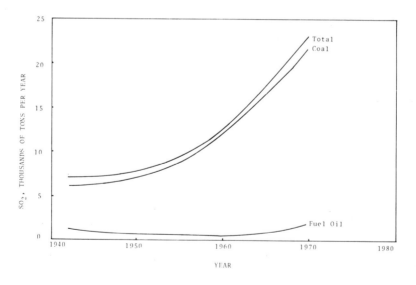

Figure 2.2   Estimated Sulfur Dioxide Emissions from
             Electric Utilities Burning Fossil Fuel
             in the U.S.

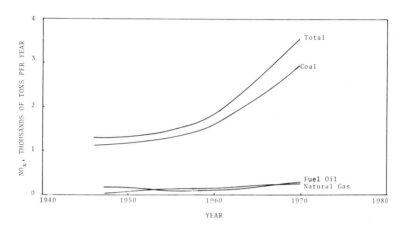

Figure 2.3   Estimated Nitrogen Oxide Emissions from
Electric Utilities Burning Fossil Fuel
in the U.S.

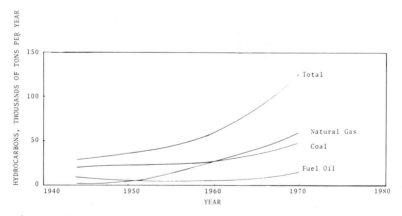

Figure 2.4   Estimated Hydrocarbon Emissions from
Electric Utilities Burning Fossil Fuel
in the U.S.

Electric power generation uses 16% of the petroleum,
19% of the natural gas, 58% of the bituminous coal
and essentially all of the water and nuclear power.
Although nuclear power is lowest on this list it
has the greatest potential future use because of

the increasing demand for power and the scarcity of
fossil fuels plus their excessive costs and air
pollution problems.  As of the end of 1970, there
were 17 operable nuclear power plants and over 100
more were under construction or being planned.  If
the advanced breeder reactors appear on schedule
in the 1980's, it is expected that nuclear power
plants will produce nearly 50% of the power by the
year 1990 and 65% of the power by 2000.

It should be noted that the increased use of
residential and commercial air conditioning has
increased power requirements such that the summer
peak energy requirements exceed the winter peak
energy requirements in all sections of the United
States except the Northwest region where the winter
peak remains greater.  This means that more fuel
must be burned to produce electrical energy in
June than in January.  This increases the air
pollution problem because it is more difficult for
the pollution to be blown away and dissipated in
the summer.

Ash and fly ash from fossil fuel power facilities can be useful.  Researchers at the University
of West Virginia have recently been able to show
that mineral wool can be successfully produced from
coal ash.  Recovered fly ash is used, for example,
in water treatment to absorb organic matter, to
extract inorganic phosphates, and to pre-condition
waste sludges before filtering.  Fly ash is also
proving valuable in asphaltic concrete, bricks,
and other building materials.  (So far only about
16% of the fly ash collected is utilized.)

## 2.4  SPACE HEATING

Household space heating produced about 9 million
tons of pollutants in 1970.  This pollution represents 0.05% by weight of fuel used (1).  By weight,
these pollutants consist of 43% $SO_2$ , 25% CO, 15%
particulate matter, 10% $NO_x$ and 7% HC.  As with
power generation, most of the space heating
facilities burn fossil fuels.  The lower $SO_2$
percentage is obtained because less coal and more
oil and gas are used in space heating.  The use of
electricity for space heating increases the demand
on the already overloaded power generation facilities.
Although this could result in a more practical
method of pollution control by requiring the control
at the centralized commercial power stations, a
different problem may result--that of concentrating

the release of pollutants into a particular localized
area. Wind ventilation is a necessary factor in the
dilution of pollutants, so this could be a serious
drawback to increasing the use of electrical power
for space heating.

## 2.5 REFUSE BURNING

Approximately 150 million tons of solid waste
must be disposed of in the United States each year.
Of this, approximately 6 million tons are disposed
of by burning. The emissions average (by weight)
20% $SO_2$, 25% HC, 15% $NO_x$, 20% CO and 20% particulate
matter. Plastics account for about 4% of this
refuse and chlorinated organic plastics such as
polyvinyl chloride produce hydrogen chloride (HCl)
and/or phosgene gases (more phosgene is produced if
combustion is poor). Mercury and mercury compounds
are also released from burning newspapers. Backyard and open type incineration are the least
desirable methods of burning refuse because this
improperly designed device is unable to provide
adequate airflow and turbulence in the combustion
zone. As a result, carcinogenic tars and other
products of incomplete combustion (e.g., CO and
particulates) are formed.

Commercial municipal incinerators, which have
been designed to meet air pollution incinerator
standards, provide adequate air and turbulence for
proper combustion, but if not properly operated
can cause severe air pollution. Properly designed
and operated modern incinerators are capable of
producing less than 100ppm CO. Control equipment
can also be used on these incinerators to further
reduce the amount of CO and $NO_x$ released. Consol-
idation of refuse burning in one large incineration
could result in a high air pollution concentration
in the vicinity of the refuse stack because of the
large quantity of releases and inadequate wind
ventilation.

## 2.6 EMISSION FACTORS

Emission factors are number values which show
the amounts of pollution emitted by weight compared
to some given basis. This basis may be the amount
of material produced, raw material used, size of
process equipment used, or plant capacity; or it
may be presented on a time and/or area basis. The
basis must be specified and the pollution controls
described if any are used.

TABLE 2.3  SUMMARY BY SOURCE OF ESTIMATED AIR POLLUTION BY WEIGHT IN 1968 (15)*

| Source | % Sulfur Oxides | % Nitrogen Oxides | % Carbon Monoxide | % Hydro-Carbons | % Particulates | Total |
|---|---|---|---|---|---|---|
| Transportation | 0.7 | 4.2 | 46.5 | 8.5 | 0.7 | 60.6 |
| Industry | 6.3 | 1.4 | 1.4 | 2.8 | 4.2 | 16.1 |
| Power Generation | 8.5 | 2.2 | 0.7 | 0.7 | 2.1 | 14.2 |
| Space Heating | 2.1 | 0.7 | 1.4 | 0.7 | 0.7 | 5.6 |
| Refuse Burning | 0.7 | 0.7 | 0.7 | 0.7 | 0.7 | 3.5 |
| TOTALS | 18.3 | 9.2 | 50.7 | 13.4 | 8.4 | 100.0 |

* not including methane

Table 2.1 includes typical emission factors that may occur for operations where no controls are used. Selected reference publications listing emission factors are given as references 3 through 16.

## 2.7 SUMMARY

Over 214 million tons of air pollutants are released into the atmosphere annually in the United States. Table 2.3 estimates by percent the major pollutants released in the U.S. in 1968 not including methane. When methane is not considered, carbon monoxide represents approximately one-half of all the pollution released into the atmosphere. Although fossil fuels generated most of the past CO in the atmosphere, most of the currently produced CO is from the internal combustion engine. The second largest pollutant is sulfur dioxide which makes up 18.3% by weight of the total pollution in the atmosphere.

## QUESTIONS FOR DISCUSSION

1. What is happening to the atmospheric content of carbon dioxide?
2. (a) List additional industrial sources of air pollution not included in Table 2.1.
   (b) Estimate their emission factors and describe how you obtained them.
3. List all of the air pollution sources within a five mile radius from where you live.
4. Attempt to give an engineering estimate of the quantity and quality of the air pollution from the sources you listed in question No. 3.

## REFERENCES

1. Rose, A.H., "Prevention and Control of Air Pollution by Process Changes or Equipment," A58-11, U.S. Public Health Service.
2. Risser, H.E., "Power and the Environment - A Potential Crisis in Energy Supply," EGN-40, Illinois State Geological Survey (1970).
3. "Compilation of Air Pollutant Emission Factors," OAP No. AP-42, Environmental Protection Agency (1972).
4. Mayer, M., "A Compilation of Air Pollutant Emission Factors for Combustion Processes, Gasoline Evaporation, and Selected Industrial Processes," U.S. Public Health Service Publication (1965).

5. "Atmospheric Emissions from Petroleum Refineries, a Guide for Measurement and Control," No. 763, U.S. Public Health Service (1960).

6. "Atmospheric Emissions from Thermal-Process Phosphoric Acid Manufacture," No. AP-48, U.S. Public Health Service (1968).

7. "Air Pollution Aspects of the Iron and Steel Industry," No. 999-AP-1, U.S. Public Health Service (1963).

8. "Atmospheric Emissions from Fuel Oil Combustion, an Inventory Guide," No. 999-AP-2, U.S. Public Health Service (1962).

9. "Air Pollution and the Craft Pulping Industry, an Annotated Bibliography," No. 999-AP-4, U.S. Public Health Service (1963).

10. "Atmospheric Emissions from Sulphuric Acid Manufacturing Processes," No. 999-AP-13, U.S. Public Health Service, (1965).

11. "Atmospheric Emissions from the Manufacture of Portland Cement," No. 999-AP-17, U.S. Public Health Service (1967).

12. "Atmospheric Emissions from Coal Combustion, an Inventory Guide," No. 999-AP-24, U.S. Public Health Service (1966).

13. "Atmospheric Emissions from Manufacturing Nitric Acid Processes," No. 999-AP-27, U.S. Public Health Service (1966).

14. "Sources of Polynuclear hydrocarbon in the Atmosphere," No. 999-AP-33, U.S. Public Health Service (1967).

15. "Emissions from Coal-Fired Power Plants, a Comprehensive Summary," No. 999-AP-35, U.S. Public Health Service, Second Edition (1969).

16. "Principle Sources of Air Pollution and Types of Pollutants," No. 1548, U.S. Public Health Service (1968).

# CHAPTER III
# POLLUTION TRANSPORT
# BY THE ATMOSPHERE

Chapter II indicates that the chain of pollution consists of source-transport-effect and some of the sources are listed in Chapter II. In this chapter we consider the atmospheric transport of pollutants from the source to the receptor (where effects will be noticed). The meteorological variables which affect the severity of an air pollution problem at a given time and location are: wind speed and direction, insolation (amount of sunlight), lapse rate (temperature variation with height), mixing depth, and precipitation. The non-meteorological factors of type and quantity of pollutants, topography and individual susceptibility must also be considered.

The atmosphere is the aerial dump for air pollution. We can be thankful that the atmosphere cleanses itself (up to a point) by natural phenomena. Atmospheric dilution occurs when the wind moves because of wind circulation or atmospheric turbulence caused by local sun intensity (insolation). Pollution is removed from the atmosphere by precipitation (rain,...) and by other reactions (both physical and chemical) as well as by gravitational fallout. A good rainstorm, for example, often scrubs the air clean and reactivates the atmosphere. Washed-out air pollution can be water pollution and must be considered as such.

The atmospheric removal mechanisms of photochemical reactions, photosynthesis and physical and chemical adsorption and absorption will be discussed in later chapters. At this time, it is sufficient to recognize that the air holds or transports the pollutants to sites where these reactions can occur.

## 3.1  WIND

Pollutants are dissipated in the atmosphere by both horizontal and vertical movement of the wind.

In general, the greater the wind velocity, the
greater the dilution.   Turbulence is a stirring
action of the wind, and therefore, is related to the
vertical movement of the wind.   Wind moving across
the surface of the earth is a fluid moving along the
surface of a containing structure.   At the surface,
the molecules of the fluid are not in motion.   The
drag against the surface of the earth causes the
wind near the surface to blow at very low speeds.
The wind speed increases very rapidly with height up
to about 10 meters.   Above this, it continues to
increase in speed with height but at a much slower
rate.
        Figure 3.1 shows how the wind may typically
increase from zero miles per hour at the surface of
the earth to almost 75 miles per hour at an altitude
of 40,000 feet.   The wind speed remains relatively
constant from ten to 100 meters   (33 to 330 feet).
Figure 3.1 also shows that the wind direction rotates
towards the right (looking downward) with increasing
height.

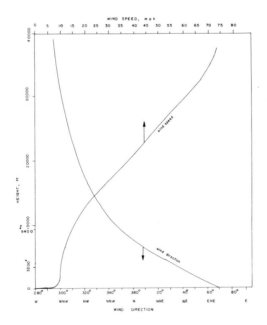

Figure 3.1   Wind Speed (Carbondale, Illinois--
            Noon, October 23, 1969)

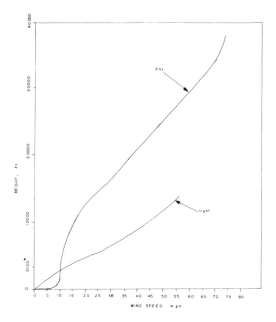

*3200' is the mean MAXIMUM MIXING DEPTH(MMD) for this location in October.

## Figure 3.2  Wind Speed Typical Variation
Day--Night

Figure 3.2 shows typical daily and nightly wind speeds and the variation that can be expected between day and night times.  Table 3.1 is a list of terms used to describe wind effects.

## 3.2  ATMOSPHERIC DIFFUSION

Atmospheric diffusion is the movement of large parcels of air from one point to another.  In this sense, it differs from the normal engineering concept of diffusion which usually is assumed to mean molecular diffusion.  Although molecules are diffusing in the atmosphere, the molecular diffusion in the atmosphere is usually negligible compared to the atmospheric diffusion of parcels of air.  Atmospheric diffusion can be likened to the movement of gas filled balloons in the atmosphere.

The travel of helium (He) filled balloons closely approximates the movement of pollutants in the atmosphere because of several factors.  First, the weight of the balloon rubber and the decreasing

TABLE 3.1   TERMS USED TO DESCRIBE WIND EFFECTS

| Terms Used in Official Forecasts | Miles Per Hour | Wind Effects Observed on Land |
|---|---|---|
| Light | 1-3 | Calm; smoke rises vertically. Direction of wind shown by smoke drift but not by wind vanes. |
| | 4-7 | Wind felt on face; leaves rustle.  Ordinary vane moved by wind. |
| Gentle | 8-12 | Leaves and small twigs in constant motion.  Wind extends light flag. |
| Moderate | 13-18 | Raises dust and loose paper. Small branches are moved. |
| Fresh | 19-24 | Small trees in leaf begin to sway. |
| Strong | 25-31 | Large branches in motion; whistling heard in telegraph wires.  Umbrellas used with difficulty. |
| | 32-38 | Whole trees in motion. Inconvenience felt in walking against wind. |
| Gale | 39-46 | Breaks twigs off trees; generally impedes progress. |
| | 47-54 | Slight structural damage occurs, such as shingles blowing off. |
| Whole gale | 55-63 | Seldom experienced inland. Trees uprooted, considerable structural damage occurs. |
| | 64-74 | Very rarely experienced but accompanied by widespread damage. |
| Hurricane | 75+ | Very rarely experienced; accompanied by widespread damage. |

density of air restrict the balloon rise to a height
that compares to the mean mixing depth which is the
limit of pollution diffusion (see Section 3.4).
Second, the small He molecules leak through the
rubber pores within 4 to 6 hours, restricting
balloon rise and causing the balloon to fall by the
time pollutants in the atmosphere would be diluted
below perception or removed by natural events. To
study this phenomena, one thousand helium filled
balloons were released at the Southern Illinois
University in Carbondale, Illinois, on October 23,
1969. Notes were attached to the balloons
requesting finders to return information as to
where the balloon landed so that a path-of-flight
map could be constructed. Figure 3.1 shows the
wind velocity at the time of balloon release.
Figure 3.3 shows the diffusion path of the balloons.
It can be noted that the low-flying balloons
traveling on the northeast winds went 70 miles to
the southwest. The highest balloons rose to almost
10,000 feet in height and traveled with the northwest winds 90 miles to the southeast. (This height
is about 3 times the average mean mixing depth, but
it was a clear, sunny day with much atmospheric
turbulence.)

Pollution released into the atmosphere would have
traveled similarly. This shows that Carbondale,
which lies 90 miles southeast of St. Louis, received
pollution from metropolitan St. Louis on that
particular day. In the case of pollution traveling
from St. Louis to Carbondale, the diffusion is
such that only about 1/25 of the pollution actually
arrives at Carbondale.

## 3.3 LAPSE RATE

The adiabatic lapse rate is the rate of cooling
with lifting (or heating upon descent) of a parcel
of air with no heat exchange. A parcel of dry air
expands upon rising. When air expands, it cools at
the adiabatic lapse rate if no heat exchange occurs.
If the surrounding air is cooler, the parcel will
continue to rise and, therefore, unstable conditions
or strong lapse rates exist. The adiabatic lapse
rate equals $-1^{\circ}C$ per 100 meters (or $-5.4^{\circ}F$ per
thousand feet).

Lapse rate is defined as equal to the adiabatic
lapse rate minus the temperature lapse rate:

$$\text{Lapse rate} = L_a - L \qquad (3.1)$$

where:   $L_a$ = -1$^\circ$C/100 meters

$$L = -\frac{\partial T}{\partial z} = -\frac{\text{change in temperature}}{\text{change in height in}}$$
$$\text{hundreds of meters}$$

Various lapse rates are shown in Figure 3.4. A lapse rate equal to or more negative than -2$^\circ$C/100 meters is a strong lapse rate and is favorable for the removal of air pollution at a given location. A strong lapse rate indicates unstable conditions and therefore, good air mixing. More positive lapse rates indicate stable air, which means air pollution will not be removed because there is not sufficient atmospheric turbulence or mixing. Inversion of the temperature lapse rate (from positive to negative) resulting in stable atmospheric conditions is known as an "inversion." Pollution concentrations can become very high during inversions. A lapse rate with a value of zero indicates neutral conditions, meaning that we have poor mixing, due only to pure mechanical turbulence created by the blowing of the wind. Neutral conditions are also unfavorable for removing pollutants from the atmosphere.

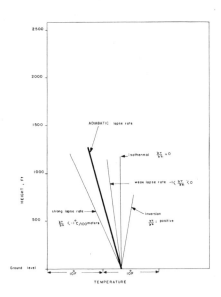

Figure 3.4  Atmospheric Lapse Rates

Unstable conditions, that is strong lapse rate,
occurs when the surrounding air is cooler and the
parcel of rising air continues to rise.  If the
temperature decreases with height more rapidly than
the adiabatic lapse rate, then the atmosphere is
unstable.  This is desirable for good air pollution
dilution in the atmosphere.  Under normal conditions,
the temperature of the air closer to the surface of
the earth is warmer during the day time than at night
time.  The temperature of the air decreases with
height during the day on sunny days, creating unstable
air and good pollution dispersion.  During the
night, the ground cools and the air closer to the
ground becomes cooler so that the temperature
increases with height, creating the stable inversion
conditions during the night time, early morning
hours, or at times when the sun rays are reaching
the earth with low intensity.
  Figure 3.5 shows a typical day-night type
temperature lapse variation.  A temperature lapse
greater than adiabatic, e.g., the unstable day
conditions shown in Figure 3.5 (a), will produce good
vertical diffusion and cause a plume to appear as in
Figure 3.5 (b).  If the temperature lapse is less
than adiabatic (approaching neutral conditions), the
plume will diffuse less vertically and look more like
Figure 3.5 (c).  A night type temperature profile
as shown in Figure 3.5 (a), with an inversion below
and strong lapse above, produces a plume that looks
like Figure 3.5 (d).  If the slope of the night or
stable day temperature profile of Figure 3.5 *remained*
positive (as for the lower portion of that curve),
inversion conditions would exist and cause the plume
to appear as in Figure 3.5 (e).  The final type of
plume, shown in Figure 3.5 (f), would occur on a
stable day when there is an inversion in the lapse
profile.  This inversion would be occurring at a
height represented by the top section of the plume.

## 3.4  MIXING DEPTHS

  The mean maximum mixing depth (MMD) is the height
to which the unstable air mixes.  In the absence of
radiosonde observations, the maximum mixing depth
may be estimated as being the height of the bottom
of the low altitude cloud layer.  The maximum mixing
depth varies during the day as well as varies from
season to season.  Variations also exist dependent
upon the topographical features.  Vertical dispersion
of pollutants is limited by the ground and the MMD,

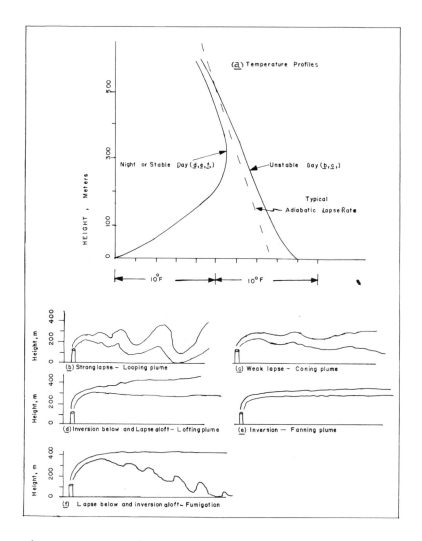

Figure 3.5   Typical Day-Night Temperature
Variations and Results on Vertical
Mixing

therefore, mixing depths are essential in estimating
the amount of vertical diffusion of pollution in the
atmosphere.   Figure 3.6 shows the MMD for the United
States (2).   MMD are lowest in December to January,
when monthly values of 400 meters are common (although

they range from less than 200 to over 1,200 meters).
The MMD steadily rise to high in May and June (common
values are 1,500 meters with a range from less than
400 to over 3,600 meters), then steadily fall to the
December-January lows.

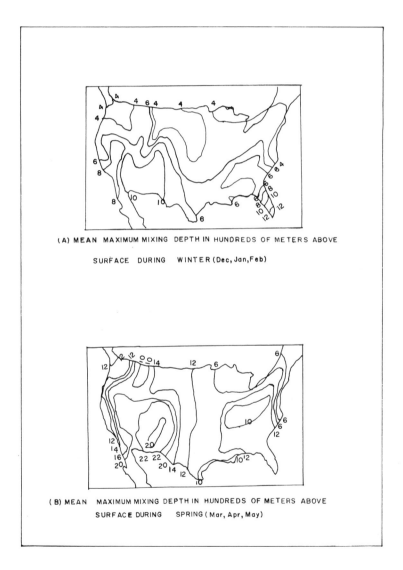

(A) MEAN MAXIMUM MIXING DEPTH IN HUNDREDS OF METERS ABOVE

SURFACE DURING WINTER (Dec, Jan, Feb)

(B) MEAN MAXIMUM MIXING DEPTH IN HUNDREDS OF METERS ABOVE

SURFACE DURING SPRING (Mar, Apr, May)

Figure 3.6a U.S. Mean Maximum Mixing Depth (MMD)(2)

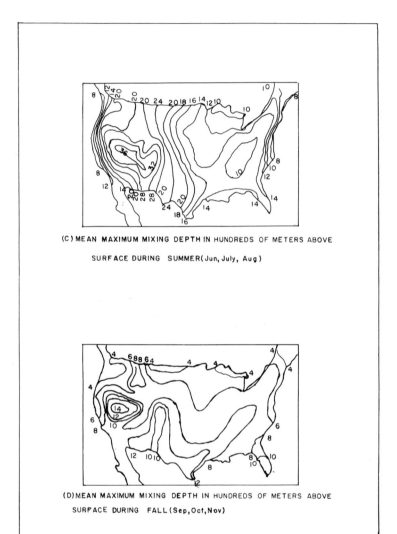

(C) MEAN MAXIMUM MIXING DEPTH IN HUNDREDS OF METERS ABOVE

SURFACE DURING SUMMER(Jun, July, Aug)

(D)MEAN MAXIMUM MIXING DEPTH IN HUNDREDS OF METERS ABOVE

SURFACE DURING FALL(Sep,Oct,Nov)

Figure 3.6b   U.S. Mean Maximum Mixing Depth (MMD)(2)

3.5   INVERSIONS

     Section 3.3 shows that inversions occur when the temperature increases with height. As stated, this occurs frequently during the night or early morning hours. In addition, inversions occur more frequently

during the fall of the year. Inversions are likened
to putting a lid on a particular locale so that no
pollution escapes by vertical diffusion. Low wind
speeds (equal or less than 7 mph) usually accompany
inversions, so there is very little horizontal dispersion of pollutants, also. Inversion temperatures
are usually limited to the first 500 meters and this,
therefore, is the maximum inversion height.
Inversion heights are usually much lower than mean
maximum mixing depth heights. Inversion heights
are usually established using weather bureau
radiosonde observations. Sketches of what may
happen to plumes released from elevated sources
during inversions are shown in Figure 3.5.

Figure 3.7 shows inversion frequency in the
United States as percent of total hours for the
different seasons (3). Note that most of these
inversions will occur during the nighttime hours.
Inversions are at a maximum during the fall, when
it is common to have inverted temperature lapse
rates over 40% of the total hours. The season for
minimum inversions is spring, when inversions are
present less than 30% of the total time.

## 3.6 SOLAR RADIATION AND WIND CIRCULATION

Most of the energy radiated from the sun travels
in wave lengths from 0.1 to 30 microns. The solar
radiation constant is approximately two Langleys
per minute at the top of the atmosphere. (Langley
is a unit of solar radiation equivalent to 1 g cal
per $cm^2$ of irradiated surface.) The atmosphere
absorbs about 19% of this radiation and the clouds
can reflect up to 25% of the total radiation.
Radiation varies with latitude. Figure 3.8 shows
both the solar radiation absorbed by the earth and
the atmosphere, as well as the long wave radiation
leaving the earth, with respect to latitude. This
figure shows that more radiation is absorbed at the
equator than at the poles. Also, more energy is
radiated from the earth at the poles than is received
at the poles. If there were no transfer of heat
toward the poles, the equatorial regions would
continue to heat up and the polar regions would
continue to cool. This thermal driving force just
mentioned is the main cause of atmospheric movement
on the surface of the earth. At the equator, warm
air rises and travels toward the poles. As it
travels toward the poles, the air cools and sinks.
In general, the air then returns toward the equator

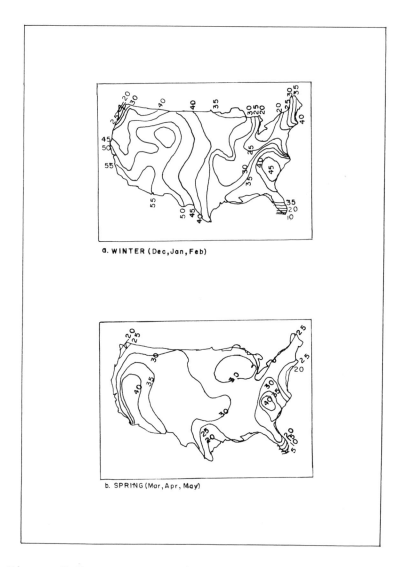

Figure 3.7   U.S. Inversion Frequency (Percent of
             total hours) (3)

as surface wind, being warmed as it travels toward
the equator.  This produces the general movement of
air from the equator to the poles at upper altitudes
and from the poles to the equator at lower altitudes.

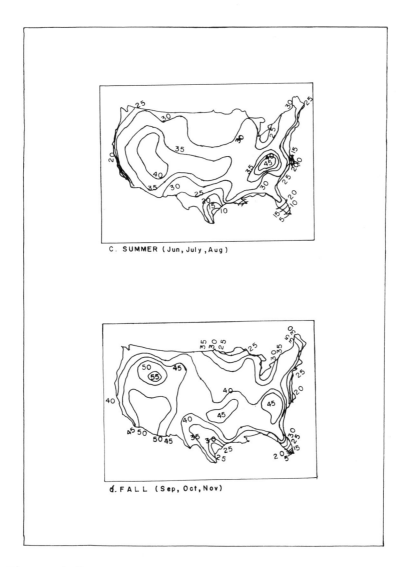

Figure 3.7    U.S. Inversion Frequency (Percent of
              total hours)(3)

The earth rotates on its axis toward the east.
The combination of air movement and earth rotation
creates in the Northern Hemisphere jet streams from
the tropics to the poles which are westerly (from
the west).  The surface winds in the area from the

equator to about $30^\circ$ to $35^\circ$ latitude are easterly, while surface winds in the area from the North Pole to about $30^\circ$ to $35^\circ$ latitude are westerly.  Cyclonic mixing at middle latitudes distributes the heat differences by creating atmospheric unstability. Cyclones are areas of low pressure that have large scale beneficial influences on air pollution dispersion.  Anti-cyclones are high pressure areas and are unfavorable for removal of air pollution in the atmosphere.  Cyclones and anti-cyclones migrate over the surface of the earth and cause day-to-day weather changes.  Anti-cyclonic areas with their clear skies and light winds are conducive to formation of nocturnal inversions.

The resultant radiation of solar energy received by the earth and the reradiation of this energy creates two cold traps in the atmosphere.  Figure 3.9 shows these cold traps as a function of extended altitudes.  The first cold trap exists in the chemosphere, and the second occurs in the ionosphere,

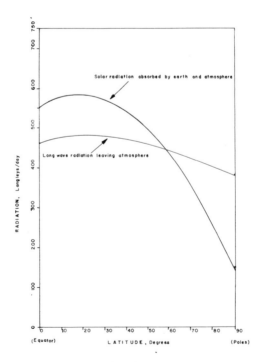

Figure 3.8   Solar Radiation

which is the region where ionic gases are formed by solar radiation. The troposphere (meaning to change) is the area where most air pollution exists. The stratosphere means to smooth out and contains very little pollution.

In addition to the thermal gradient driving force and the effect of the earth's rotation, there is a pressure gradient force which attempts to move air from high pressure areas (anti-cyclonic) to low pressure areas (cyclones). Wind motion is therefore in three dimensions, although we usually only con- sider the horizontal component. The coriolis force results because the air attempts to move in a straight line (inertial force) while the earth rotates. This force is at right angles to the wind velocity in the northern hemisphere and at left angles to the wind velocity in the southern hemisphere. The coriolis force is proportional to wind velocity and decreases with latitude.

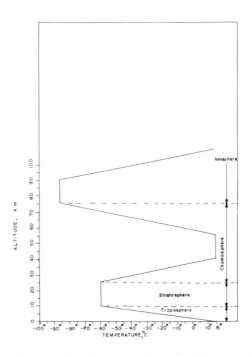

Figure 3.9    Typical Atmospheric Temperatures
             (Cold Traps)

In summary, the forces which result in wind circulation are:  inertial force resulting from the thermal gradient, the drag force (which is opposite in direction to the wind inertial force), the pressure force and the coriolis force.

## 3.7   PRECIPITATION

Pollutants can be washed out of the air by the natural scrubbing action of rain, snow and all other forms of precipitation as it falls to the ground. Gases that are soluble in water are removed by absorption.  Particulate matter is removed by adsorption when the particles stick to the precipitation after being impacted by it.  Scrubbing action is further discussed under pollution control in Sections 9.1, 10.3 and 10.4.

Not only can precipitation affect the atmospheric pollution, but the pollution can affect the precipitation.  Peterson (1) summarizes data that appears to indicate that the presence of particulates in urban atmosphere causes increased amounts of precipitation due to seeding action.  Data obtained indicate that some central United States urban areas have 5 to 31% more annual precipitation than the surrounding rural areas, and have up to 38% more thunderstorm days annually.  There is another theory that cloud superseeding also occurs because of the pollution, reducing the amount of precipitation and causing drought.  This theory is proposed to account for the dryness in some areas of the United States.

## 3.8   TOPOGRAPHIC INFLUENCES

Topography can very seriously affect local atmospheric conditions.  Most important of these are valleys, shorelines and hills.  Most of the infamous air pollution episodes occurred in locations with adverse topographical conditions.  The following three events are examples of this.  (1)  During the September 1955 heat and smog siege in Los Angeles County, 1,000 deaths over normal occurred.  Los Angeles has the adverse topographical feature of a line with mountains rising immediately behind the shoreline.  (2)  In Donora (south of Pittsburgh, Pennsylvania), when an air pollution episode was created because ten times the normal sulfur dioxide concentration existed in 1948, twenty persons died and 5,910 became ill during a three day period.

This amounted to half the population. Ten years
later, people in Donora were still dying prematurely.
Donora is located in a valley between two mountain
ranges. (3) London has infamously poor topographical features and, as a result, has had numerous
air pollution episodes. During the 1952 London fog
episode, 4,000 people died plus a number of prize
cattle being shown at a local exhibit died as a
result of overexposure to high $SO_2$ and particulate
concentrations.

Valley effects occur due to the channeling of
winds. Valleys tend to make the wind flow in the
general direction of the valley axis. Slope winds
occur in valleys in the evenings when the air near
the ground is cooled by the ground. Figure 3.10 (a)
shows that the air at location "A" cools quickly in
the evening and begins to flow down the valley toward
point "B." In Figure 3.10 (b), the cold air, which
has moved down the slopes, begins to collect in the
valleys until it becomes deep enough to flow down
the valley creating valley winds.

Any air pollution released on the slopes or in
the bottom of a valley will stay in the valley if
an inversion situation also exists. Figure 3.10 (c)
shows both slope winds and valley winds moving
downhill. The onset of the valley wind usually
lags several hours after the onset of the slope
winds. In reverse, on a clear morning with a light
wind, heating can cause up-slope and up-valley winds.

Shoreline winds are created because of the
differences in heating rates of the earth and the
water. Under the same amount of sunshine, land
absorbs heat faster than water. Therefore, on a
sunny day, sea breezes are created which come from
the water to the land. Land also cools faster than
water creating, on cloudy days or at nighttime,
land breezes. Land breezes are usually lower in
velocity and shallower than sea breezes. Figure 3.11
shows how a sea breeze can create undesirable
downwash from a stack located near the shoreline.

Hills can cause a varying degree of influence
on air pollution removal. A smooth hill causes the
least effect on the flow of air. Rough hills,
conversely, cause turbulent eddies and good mixing
which promote improved air pollution removal. If
a mountain is sufficiently high, it will cause the
wind to deflect and flow parallel to the mountain
and not over the mountain. This condition occurs
in Los Angeles where the cool, moist ocean air does
not have sufficient bouyancy to rise and blow the
pollution over the surrounding mountains.

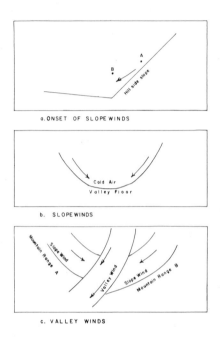

a. ONSET OF SLOPE WINDS

b. SLOPE WINDS

c. VALLEY WINDS

Figure 3.10   Mountain and Valley Winds

3.9   METEOROLOGICAL ROSES

A wind rose is a diagram designed to show the
distribution of wind direction experience at a given
location over a considerable period of time. The
rose is simply a pictorial graph showing prevailing
wind direction. The lines or bars from which a rose
may be constructed shows the direction from which
the wind is blowing. There are sixteen total wind
directions. The eight principal wind directions are:
N, NE, E, SE, etc. and the eight secondary wind
directions are: NNE, ENE, ESE, SSE, etc. (e.g., see
Figure 3.28, where lengths of lines are proportional
to frequency of wind in the indicated direction).
It may be necessary to remove the bias in the
data before a wind rose is constructed. Bias
frequently exists when data are obtained over short
sampling periods. Bias is checked for by summing
the principal wind direction frequencies and
comparing with the sum of the secondary direction
frequencies. Bias exists if one of these exceeds

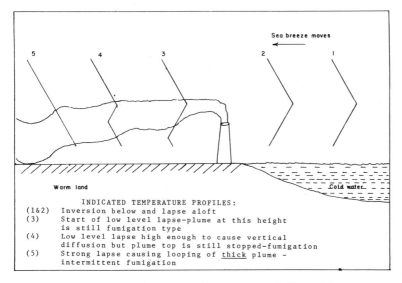

Figure 3.11   Daytime Sea Breeze and Results on a
Plume of Pollution

the other by 15%, and it can be removed by the
following procedure:

For $N_e > N_o$

where:   $N_e$ equals the sum of the principal
direction frequencies

$N_o$ equals the sum of the secondary
direction frequencies

*Step 1:*   Subtract D from each primary
direction--

$$D = n_e \left( \frac{N_e - N_o}{2N_e} \right)$$   (3.2)

where:   $n_e$ = frequency for any one particular
principal direction and wind speed group.

*Step 2:*   Add E to each secondary direction--

$$E = n_o \left( \frac{N_e - N_o}{2N_o} \right)$$   (3.3)

where:   $n_0$ = frequency for any one particular
          secondary direction and wind speed group.

When $N_0 > N_e$, add in Step 1 and subtract in Step 2.

   The next step in construction of the wind rose
is to distribute the calms.  During the taking of
data, there are usually periods of calm when no
wind blows.  If this occurs, these non-directional,
zero wind speed data points must be included in the
data and assigned a specific direction.  It is
desirable to distribute the calms and the lowest
wind speeds (up to 7 mph) among the sixteen
directions in order to have a more representative
sample of light winds.  The number of calms (F)
assigned to a particular direction is calculated
by the following procedure:

$$F = n_w \left( \frac{N_c}{N_w} \right) \qquad\qquad (3.4)$$

where:   $N_c$ = total number of calms

          $N_w$ = total frequency of winds in the
              0-7 mph range (not including calms)

          $n_w$ = frequency of the 0-7 mph winds in
              one particular direction.

See problem 3.4 for a sample of the above procedure
for constructing wind roses.

## 3.10   INTRODUCTION FOR DIFFUSION CALCULATIONS

   A plume of pollution released into the atmosphere
begins to move downwind with the prevailing wind.
As it moves, it diffuses both vertically and
horizontally from a line representing the center
line.  The rate at which diffusion occurs depends
on:  (1) wind speed, (2) insolation (amount of
sunshine), (3) other factors which cause disturbance
and turbulence in the air (such as both man-made and
natural topographical obstacles), (4) effective stack
height of release, (5) source strength, (6) lapse
rate, and (7) mixing height.  The distribution of
plume concentration around the center line is usually
considered to be Gaussian, with the distribution
values being expressed as deviations.

Statistics show us that we can calculate deviations by the following procedure:

Averages are obtained by dividing the sum of the individual values by the frequency

$$\bar{\alpha} = \frac{\Sigma \alpha}{n} \qquad (3.5)$$

Deviation from the mean is equal to the value minus the average value

$$\alpha' = \alpha - \bar{\alpha} \qquad (3.6)$$

The standard deviation is then equal to the square root of the sum of the deviations squared divided by the frequency

$$\sigma = \sqrt{\frac{\Sigma \ (\alpha')^2}{n}} \qquad (3.7)$$

Figure 3.12 shows a pictorial representation of a plume which is diffusing vertically with increasing downwind distance. Note that the effective height of the center line of the plume is 200 meters. The stack obviously is not 200 meters high, but the plume has some lift (plume rise) making the center line of the plume above the top of the stack. At a distance of 100 meters downwind, this plume has diffused 10 meters in a vertical direction. 150 meters downwind, the plume has diffused 14 meters in a vertical direction from the center line of the plume.

A cross section of Figure 3.12 is shown in Figure 3.13 at a distance of 100 meters downwind. It shows that the plume is elliptical in shape. Deviation in the horizontal direction ($\sigma_y$) is usually greater than deviation in the vertical direction ($\sigma_z$). Horizontal deviation is a function of the downwind turbulence where $\sigma_z$ is more a function of the mixing turbulence due to the atmospheric insolution and lapse rate. Figure 3.14 shows a schematic of plume diffusion in both vertical and horizontal directions as a function of Gaussian distribution. With this system, $\sigma_y$ and $\sigma_z$ signify that approximately two-thirds of the pollution will be within the values given for $\sigma_y$ and $\sigma_z$. The characteristics of Gaussian or log normal distribution are shown by the bell shaped curve of Figure 3.15, where the equation of the ordinate is:

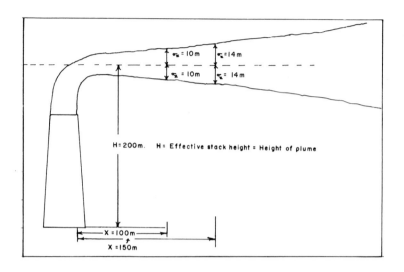

Figure 3.12   Plume Diffusion
              (Note:  Although the $\sigma_z$ are shown to
              include the entire plume, they actually
              only include 67% of the distance.)

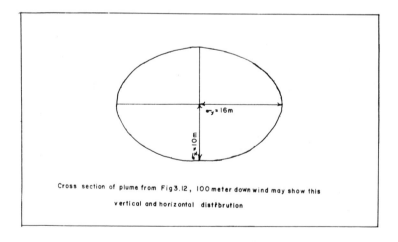

Figure 3.13   Plume Cross Section

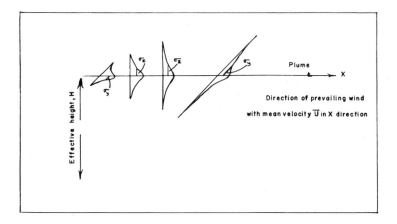

Figure 3.14    Schematic of a Plume Showing Gaussian
Distribution

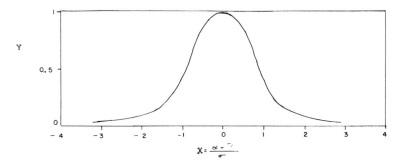

Figure 3.15    Gaussian Distribution Curve

$$Y = \frac{1}{\sqrt{2\pi}\,\sigma} \quad \exp\left[ -\frac{1}{2}\left(\frac{\alpha-\overline{\alpha}}{\sigma}\right)^2\right]$$

$$X = \frac{\alpha-\overline{\alpha}}{\sigma}$$

(Note that the term "exp" means the natural
logarithm e raised to the quantity following exp.)

The area under the curve of Figure 3.15 from - ∞
to any value of X is equal to the = ∫ Ydα = ∫ σ Y dX

$$X = \int\limits_{-\infty}^{\left(\frac{\alpha-\overline{\alpha}}{\sigma}\right)} \frac{1}{\sqrt{2\pi}} \exp\left[-\frac{1}{2}\left(\frac{\alpha-\overline{\alpha}}{\sigma}\right)^2\right] dX \qquad (3.8)$$

Figure 3.16 shows that a plot of the area under a
normal distribution curve gives a straight line on
Cartesian probability scales when plotted against
values of X. (The probability scale is discussed in
Section 7.1.3).

Diffusion in the vertical direction will, at
some downwind distance, produce a reflection condition
as shown in Figure 3.17. Beyond this point of
reflection, the diffusion cannot continue in the
vertical direction but is turned back upon itself
resulting in a doubling of the concentration. This
condition is accounted for in the diffusion equations
given in Section 3.11. Obviously, pollution released
at ground level has a reflection distance of zero

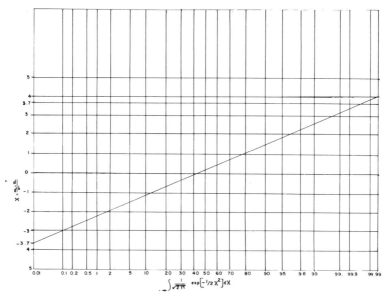

Figure 3.16   Probability Plot (X vs Area under
              Distribution Curve from -∞ to any
              Value X)

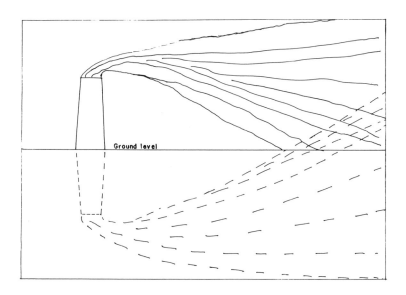

Figure 3.17    Ground Level Concentration Due to
                Reflected Source

if there is no effective rise of the plume.  One
may envision the reflected concentration as coming
from an imaginary second source, such as shown by
the dotted lines in Figure 3.17.
     There are several types of diffusion equations
which can be used.  The development of the Sutton
equation, modified with Pasquill's diffusion
deviation parameters, and using the empirical data
of Gifford for diffusion estimate values appears to
give the best results at this time.  Therefore, this
is the method presented in Section 3.11.  It should
also be noted that several other methods are
currently being developed.  Sutton and Taylor have
developed diffusion equations using 1 point diffusion
theories.  This point is obtained when two line
sources intersect, giving a point source.  The
results of both theories terminate in the same type
of final equation.  Taylor developed this equation
using K theory.  K is eddy diffusivity, which varies
with height of source, type of day, time of day,
type of terrain and wind speed.  The factors of
time of day and type of terrain are not included in
the Sutton procedure used in Section 3.11.  These
factors could produce valuable additional improvements in diffusion calculations when the procedure
has been more fully developed.

Taylor's momentum force equation for force in the downwind direction is

$$-F_{zu} = \rho K_z = \frac{\partial u}{\partial z}$$

(3.9)

where:  $F_{zu}$ = the momentum force in the downwind direction acting on the air in the z plane

$\rho$ = density of the air

$K_z$ = the eddy diffusivity in the z plane

$u$ = wind velocity

$z$ = height

This equation is proportional to $\tau_{zu}$ where $\tau_{zu}$ is the downward sheer stress momentum in the downwind direction in the z plane. Eddy diffusivity is proportional to height:

$$\frac{K_z}{K_{z_1}} = \left(\frac{z}{z_1}\right)^n$$

(3.10)

In the K theory, n equals one, therefore, K varies linearly with the height.

The K theory assumes:  Gaussian distribution, wind constant with height and no diffusion in the x direction.  The resultant transport equation of continuity at steady state gives:

$$Q = \iiint x \, dx dy dz$$

(3.11)

where:  Q = source strength

The resulting diffusion equation is similar to the Sutton equation (Equation 3.12) used in this work.

3.11   ATMOSPHERIC DIFFUSION CALCULATIONS

The equations presented in this section are variations of Sutton's (4) equation as modified with Pasquill's (5) diffusion parameters.  The tables of diffusion deviation estimates are empirical data as described by Gifford (6).

This diffusion calculation procedure considers source height, and wind speed.  It assumes Gaussian distribution and no diffusion in the x direction. It also includes a crude procedure for estimating stability classes depending on type of day and time of day as indicated in Table 3.2--Key to Stability Categories. Using Table 3.2 to select the proper stability key letter, the appropriate curves of Figures 3.17 through 3.26 can be used to obtain desired diffusion deviation estimates for use in the diffusion equations.

It is important to note that the accuracy of these calculations can vary up to 2 to 5 times the actual measured concentrations. The ability of the observer to correctly establish the stability key letter is most important--calculated results are accurate and worthwhile only if this is done carefully.

### 3.11.1 DIFFUSION EQUATIONS FOR GASEOUS POLLUTANTS

*POINT SOURCES*

General Equation·-The general formula presented below can be used to calculate downwind concentration at any point x, y, z with respect to the location of the emission source. In the denominator, one $2\pi$ is obtained for every direction of Gaussian distribution. Remembering that there is no distribution in the x direction, this leaves only y and z (horizontal and vertical) diffusion.

$$C_{(x,y,z)} = \frac{Q}{2\pi \bar{u} \sigma_y \sigma_z} \exp\left[-\frac{y^2}{2\sigma_y^2}\right] \left\{ \exp\left[-\frac{(z-H)^2}{2\sigma_z^2}\right] \right.$$

$$\left. + \exp\left[-\frac{(z+H)^2}{2\sigma_z^2}\right] \right\} \qquad (3.12)$$

where: $C_{(x,y,z)}$ = concentration downwind from source at position x,y,z, $g/m^3$

$Q$ = source strength when pollution released, g/sec

$\bar{u}$ = mean wind speed, meters/sec

$H$ = effective plume height, meters

$x$ = distance downwind from source, meters

$y$ = distance horizontally from plume center line, meters

$z$ = distance above ground, meters

$\sigma_y$ = horizontal deviation, meters

$\sigma_z$ = vertical deviation, meters

Values for the terms in equation 3.12 are normally
all obtained by direct measurement and calculation
except for the values of $\sigma_y$ and $\sigma_z$ which are
obtained from Figures 3.18 through 3.26 after
having determined the proper stability key letter
from Table 3.2.

Downwind on Ground (Ground Source)--Equation 3.12
can be simplified for various conditions.  The
concentration at ground level on the center line of
the plume when the pollution is emitted from a
ground source is obtained by:

$$C_{(x,0,0)} = \frac{Q}{\pi \bar{u} \sigma_y \sigma_z} \qquad (3.13)$$

The concentration of pollution at ground level
and on the center line of the plume from a ground
source during an inversion fumigation (which means
there is no diffusion in the vertical direction and
therefore, no $\sigma_z$) is calculated from:

$$C_{(x,0,0)} = \frac{Q}{\sqrt{2\pi} \bar{u} \sigma_y h} \qquad (3.14)$$

where:  h = ht. of uniformly mixed
inversion layer, meters

Concentration at any point xyz from a ground
source can be obtained using:

$$C_{(x,y,z)} = \frac{Q}{\pi \bar{u} \sigma_y \sigma_z} \exp -[\frac{y^2}{2\sigma_y^2} + \frac{z^2}{2\sigma_z^2}]$$
$$\qquad (3.15)$$

Downwind on Ground (Elevated Source)--Equation 3.12
can be arranged to give the ground concentration on
the center line from an elevated source by:

$$C_{(x,0,0)} = \frac{Q}{\pi \bar{u} \sigma_y \sigma_z} \exp [-\frac{H^2}{2\sigma_z^2}] \qquad (3.16)$$

The *most frequently used version* of equation
3.12 for obtaining concentration on the ground
from an elevated source at any given crosswind
distance is:

$$C_{(x,y,0)} = \frac{Q}{\pi \bar{u} \sigma_y \sigma_z} \exp -[\frac{y^2}{2\sigma_y^2} + \frac{H^2}{2\sigma_z^2}] \qquad (3.17)$$

TABLE 3.2 KEY TO STABILITY CATEGORIES

| Surface Wind Speed at Ten Meters Height m/sec | INSOLATION STABILITY CLASSES | | | | |
|---|---|---|---|---|---|
| | Day | | | Night | |
| | Strong (1) | Moderate (2) | Slight (3) | Thinly Overcast or >½ Cloud (4) | Clear to <½ Cloud |
| >2 (4.5mph) | A | A-B | B | – | – |
| 2-3 (4.5-6.7) | A-B | B | C | E | F |
| 3-5 (6.7-11) | B | B-C | C | D | E |
| 5-6 (11-13.5) | C | C-D | D | D | D |
| >6 (>13.5mph) | C | D | D | D | D |

(1) Sun>60° above horizontal; sunny summer afternoon; very convective
(2) Summer day with few broken clouds
(3) Sunny fall afternoon; summer day with broken low clouds; or summer day with sun from 15-35° with clear sky
(4) Winter day

Insolation = amount of sunshine

Class A indicates greatest amount of spreading and most unstable atmospheric conditions and Class F indicates least spreading and most stable atmospheric conditions.

The special case when concentration at any point x,y,z is desired from an elevated source when x is small so that no reflected source need be considered, use the following equation:

$$C_{(x,y,z)} = \frac{Q}{2\pi\bar{u}\sigma_y\sigma_z} \exp -[\frac{y^2}{2\sigma_y^2} + \frac{(z-H)^2}{2\sigma_z^2}] \qquad (3.18)$$

Maximum Ground Concentration (Elevated Source)--It is frequently necessary to know where the maximum ground concentration occurs from an elevated source and what its value is. The maximum concentration occurs *approximately* where:

$$\sigma_z = \frac{H}{\sqrt{2}} \qquad (3.19)$$

and the maximum concentration can be quickly obtained using:

$$C_{(x,o,o)max} = \frac{0.117\ Q}{\bar{u}\sigma_y\sigma_z} \qquad (3.20)$$

Independent values of $\sigma_y$ and $\sigma_z$ can be used from Figures 3.18 and 3.19 or the product, $\sigma_y\sigma_z$ may be used from Figure 3.25.

Downwind Concentration (Elevated Source)--The center line concentration at any height from an elevated source may be found using:

$$C_{(x,o,z)} = \frac{Q}{2\pi\bar{u}\sigma_y\sigma_z} \exp -[\frac{(z-H)^2}{2\sigma_z^2} + \frac{(z+H)^2}{2\sigma_z^2}] \qquad (3.21)$$

*INVERSIONS*

If there is an inversion limit, special precautions must be taken. At the distance $x_h$, the stable layer restricts further vertical diffusion. At the distance $2x_h$, vertical mixing has produced a completely homogenous uniform concentration. It is, therefore, necessary to know at what distance the inversion lid has limited further diffusion in the vertical direction. Calculate $x_h$ using:

$$\sigma_{z_h} = 0.47\ h \qquad (3.22)$$

where:  h = inversion height in meters

This value of $\sigma_{z_h}$ is then used to give $x_h$ from Figure 3.18. For all distances less than $2x_h$ use the standard equations already given and regular values of $\sigma_y$ and $\sigma_z$. For distances equal or greater than $2x_h$, use:

a constant value of $\sigma_z = 0.8h$, regular values of $\sigma_y$, and either equation 3.13 or 3.23 (note that when equation 3.13 is modified by this procedure, it is approximately equivalent to equation 3.14).

$$C_{(x,y,z)} = \frac{Q}{\pi \bar{u} \sigma_y \sigma_z} \exp \left[ - \frac{y^2}{2\sigma_y^2} \right] \qquad (3.23) /$$

## TURBULENT WAKES

If a building or other obstacle is in the path of a plume and creates a turbulent wake, use

$$\sigma_z' = (\sigma_z^2 + \delta^2)^{\frac{1}{2}} \qquad (3.24)$$

and

$$\sigma_y' = (\sigma_y^2 + \delta^2)^{\frac{1}{2}} \qquad (3.25)$$

where: $\delta \simeq 2$ times height of obstacle, meters

The normal equations presented before then can be used with these values of $\sigma_y'$ and $\sigma_z'$.

## NON-POINTS (AREA) SOURCES

Pollution concentration resulting from ground level emission of pollutants from several sources or from a line source such as fires can be calculated for ground level center line positions as follows:

$$C_{(x,o,o)} = \frac{Q}{\pi \bar{u} (\sigma_y^2 + \sigma_{y_o}^2)^{\frac{1}{2}} \sigma_z} \qquad (3.26)$$

where: $\sigma_{y_o} = \frac{1}{4}$ the emission width, meters

## RELEASE FROM CONFINEMENT

For the case where pollutants are released within a room and from there leak to the outside, the following equation can be used:

$$C_{(x,o,o)} = \frac{Q}{(\pi \sigma_y \sigma_z + A') \bar{u}} \qquad (3.27)$$

where:   $A' = \frac{1}{2}$ the cross section of the
building perpendicular to the
flow of the wind

## RADIATION POLLUTION

Any of the above equations can be used for
radiation pollution calculations.  It is necessary
only to use the proper equation and to use C and Q
in terms of curies instead of grams.  Radiation
decay must be accounted for when necessary.

### 3.11.2  DIFFUSION EQUATIONS FOR PARTICULATE POLLUTANTS

### POINT SOURCES (Tilted Plume Model)

Concentration of particulate matter which is
released from elevated sources can be calculated
for center line ground level concentration using:

$$C_{(x,o,o)} = \frac{PQ}{\bar{u}\sigma_y\sigma_z} \; \exp \; [-\frac{1}{2}\frac{B^2}{\sigma_x{}^2}] \qquad (3.28)$$

where:   P = wt. fraction of effluent in a
particular size range

$$B = H - \frac{v_s x}{100 \; \bar{u}}$$

$v_s$ = terminal settling velocity for the
particular size range used for P
(see Section 3.13), cm/sec.

It is important to choose small ranges of size
when determining P as $v_s$ varies considerably for
different size materials.  Classification of
particulate matter into specific size groups will
be discussed in Part II of this text under Engineering Control.  Terminal settling velocities are
discussed in Section 3.13 of this text.

### INSTANTANEOUS RELEASES

Concentrations of pollution resulting from
instantaneous puff-type releases from elevated
sources can be calculated using the following
equation (this equation is good for either radioactive or nonradioactive particulate matter):

$$D_{t(x,y,z)} = \frac{Q_t}{\pi \bar{u} \sigma_y \sigma_z} \exp \left[ - \frac{1}{2} \left( \frac{y^2}{\sigma_y^2} + \frac{H^2}{\sigma_z^2} \right) \right]$$

(3.29)

where:  $D_t$ = total dosage, g sec/m

$Q_t$ = total release, g

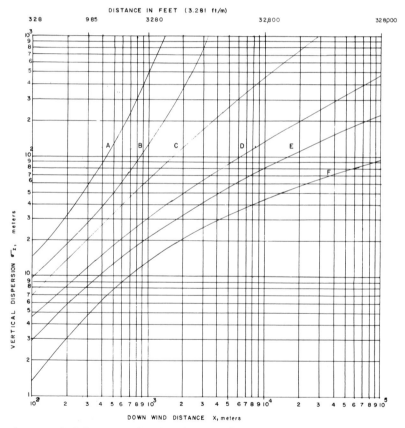

Figure 3.18  Vertical Dispersion Standard
Deviations over Open Country

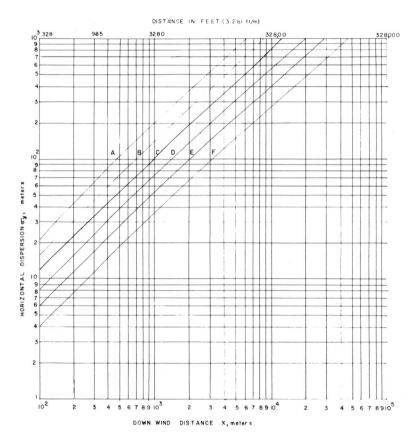

Figure 3.19   Horizontal Dispersion Standard
Deviations over Open Country

3.12   PLUME RISE

Many plume rise formulas are available, however,
most have limitations (e.g., used only for low
stacks, limited temperature differential between
stack gas and air, used only for low wind speed,
etc..).   A recent summary of different stack rise
equations is presented by Thomas et al. (7) where
they compare the plume rise equations of Holland;
Bosanquet, Carey and Holten; Davidson-Bryant; and
six other plume rise equations or modified equations.
The Holland equation is one of the older and more
reliable equations and is presented here:

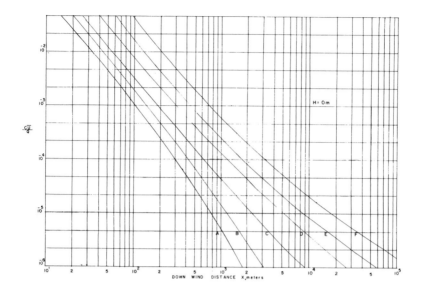

Figure 3.20   Diffusion Estimating Factors for
Elevated Emissions

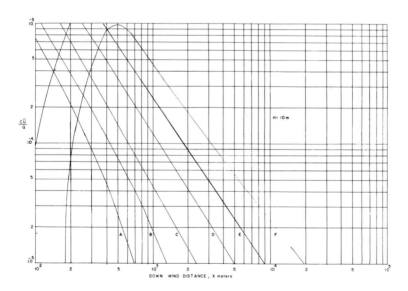

Figure 3.21   Diffusion Estimating Factors for
Elevated Emissions

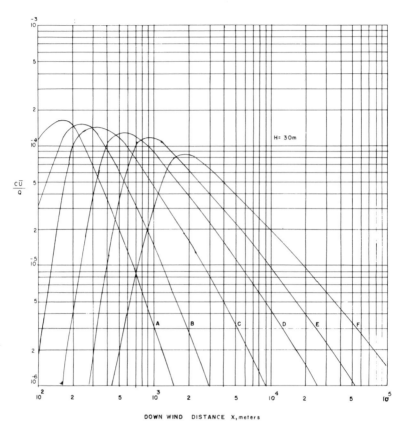

DOWN WIND  DISTANCE X, meters

Figure 3.22   Diffusion Estimating Factors for
Elevated Emissions

$$\Delta H = \frac{V_e D}{\bar{u}} (1.5 + 2.68 \times 10^{-3} \ P \ \frac{\Delta T}{T_s} D) \qquad (3.30)$$

where:  $\Delta H$ = plume rise above stack exit, meters
D = stack ID, m
$V_e$ = stack exit velocity, m/sec.
$T_s$ = stack temperature, $^{\circ}$K (or $^{\circ}$R)
$\Delta T$ = $T_s$ - ambient air temp., $^{\circ}$K (or $^{\circ}$R)
P = atmospheric pressure, mb
(1 atm = 1013 mb)

The Holland plume rise equation is correct for
neutral conditions.  The equation should be

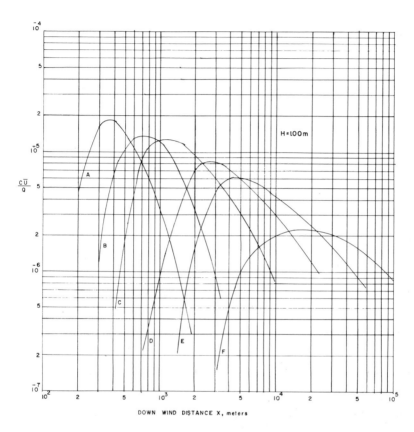

Figure 3.23  Diffusion Estimating Factors for
Elevated Emissions

multiplied by 0.8 for stable conditions and
multiplied by 1.2 for unstable conditions.  It
should also be noted that topography affects plume
rise, therefore, physical observations of the stack
and the surrounding community will be important
when making plume rise estimations.  One of the
most recent correlations of plume rise and field
observations is reported by Fay (8).

Washed stack gases can create undesirable plume
rise conditions.  During gas washing to remove
pollutants, the gases are cooled.  This can create
negative buoyance and reduce the effective stack
height to ground level (that is, H may be equal to

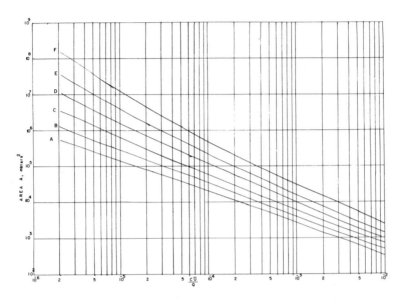

Figure 3.24   Areas of Constant Concentration
            Isopleths for Ground Level Emissions

zero). Aerodynamic downwash can also result in
effective stack heights equal to zero. This is
caused by mechanical turbulence, temperature
inversions, or when high winds traveling horizontally across the stack cause the plume to drop to
ground level. Downwash is most often noticed in
the downwind region located from two to ten times
the height of the stack.

3.13   ATMOSPHERIC PURIFICATION

Pollutants are removed from the local atmosphere
by the diffusional transport mechanism of the wind
which blows the pollution away. Pollution can be
separated out and removed from the atmosphere by
precipitation, gravitational settling, chemical
reactions, and radioactive decay. We discussed in
Section 3.7 how precipitation aids in the removal
of pollution. Particles larger than 10 microns in
size in the atmosphere easily fall out. This
gravitational settling mechanism is natural and
necessary for purifying air although it does create
a dust nuisance problem. In free fall, the particles
are acted on by the forces of gravity, aerodynamic

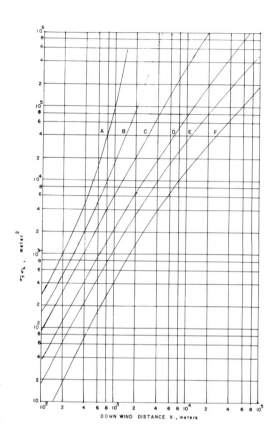

Figure 3.25 Product of Dispersion Standard
Deviations over Open Country

drag and buoyancy. By combining these forces as
is shown in Part II under Engineering Control, we
obtain Stokes' equation for terminal settling:

$$v_s = \frac{2r_p^2 g \, \rho_p}{9 \, \mu_a} \qquad (3.31)$$

where:  $v_s$ = terminal settling velocity, cm/sec
$r_p$ = particle radius, cm (1 micron =
$10^{-4}$ cm)
$\rho_p$ = particle density, g/cm³
$\mu_a$ = air viscosity = 1.8 x $10^{-4}$ g/(cm sec)
$g$ = gravitational acceleration =
980 cm/sec²

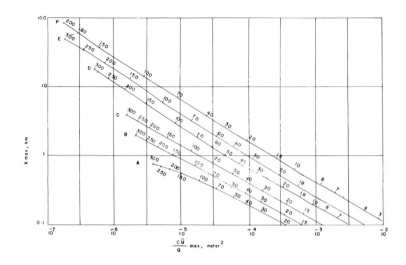

Figure 3.26   Distance of Maximum Concentration and
             Maximum C$\bar{u}$/Q as a Function of Stability
             (Curves) and Effective Height (Meters)
             of Emission (1)

Figure 3.27 is a plot showing terminal settling
velocity versus particle diameter. Very small
particles would tend to slip through the molecules
of air while settling, and therefore, a correction
factor is included to account for this in Figure
3.27. This is called the Cunningham correction
factor. Also, Equation 3.31 must be corrected for
large diameter particles--this is included in
Figure 3.27. To obtain terminal settling velocity
for any other density particle, use the correction:

$$v_{s(x)} = v_{s(2)} \frac{\rho_x}{2.0} \qquad (3.32)$$

where:  $v_{s(x)}$ = settling cm/sec for particle
                   of density $\rho_x$
        $v_{s(2)}$ = settling cm/sec from Figure 3.27
                   for particle of density = 2 g/cm$^3$

SUGGESTED READING SOURCES

The following three publications are recommended
as excellent sources of extra reading:

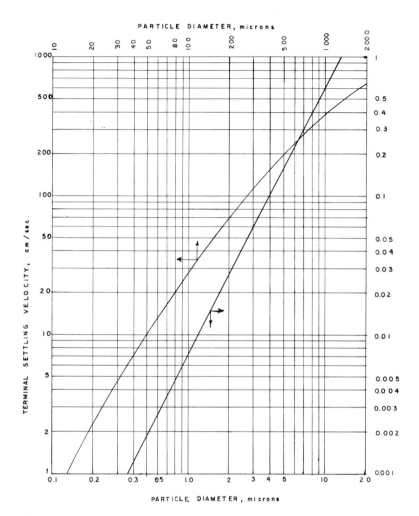

Figure 3.27  Corrected Terminal Settling Velocity
of 2 g/cm³ Spherical Particles in Air
at 25°C and 1 Atmosphere

1. Turner, D.B., "Workbook of Atmospheric
   Dispersion Estimates," United States Department
   of Health, Education and Welfare, Public Health
   Service (Revised 1969).
2. Slade, D.H. (Editor), "Meteorology and Atomic
   Energy," available as TID-24190 from Clearing
   House for Federal Scientific and Technical

Information, National Bureau of Standards, U.S.
Department of Commerce, Springfield, Virginia
22151 ($3.00) (1968).
3.   "Tall Stacks, Various Atmospheric Phenomena and
Related Aspects," U.S. Department of Health,
Education and Welfare, Public Health Service
(1969).

QUESTIONS FOR DISCUSSION

1.   Describe how a laboratory model can be made to
show when slope winds begin to blow and in what
direction.
2.   Repeat No. 1 for valley winds.
3.   How could you obtain data to measure the lapse
rate to determine the effect on air pollution
dispersal?
4.   The following are units of pressure, work and
force.  Tell which each is:  dyne, erg, bar.
5.   Give values of standard pressure in units of
atmospheres, psia, feet of water, inches of mercury,
millimeters of mercury, millibars, and torr.

PROBLEMS

3.1   Calculate the density of air at S C ($60^{\circ}F$).
3.2   A power plant emits 0.25% $SO_2$ from its stack
at a temperature of $275^{\circ}F$.  The volume of gas
emitted is 5 x $10^5$ cu ft/min.  What is the $SO_2$
emission rate in g/sec?
3.3   Air at $10^{\circ}F$ is sampled for $SO_2$ by passing it
through a bubbler.  It is warmed to $77^{\circ}F$ before flow
rates are taken.  What correction must be applied
to give values for true outside concentration?
(Neglect pressure drop.)
3.4   Table 3.3 is wind speed and direction data for
St. Louis, Missouri.  Using Table 3.4 as a guide,
finish removing the bias from this data and finish
Table 3.4  Using Table 3.5 as a guide, finish
distributing the calms from the data of Table 3.3
and finish Table 3.5.  Using the data of Tables
3.4 and 3.5, construct a debiased wind rose with
the calms distributed as started in Figure 3.28.
3.5   The measured $SO_2$ concentration is 34 $\mu g/m^3$.
Air temperature and pressure is $44^{\circ}F$ and 987 mb.
What is the volume concentration in parts per
hundred million?
3.6   A power plant burns 200 tons of coal per day
maximum on a sunny October day ($32^{\circ}F$).  The coal
contains 1% S and 11% ash.  Calculate the $SO_2$

TABLE 3.3   WIND DIRECTION AND SPEED FOR ST. LOUIS, MISSOURI (OCTOBER)

| Direction | *Frequency of Wind Speed in MPH Observed at Hourly Intervals* | | | | | |
|---|---|---|---|---|---|---|
| | *0-3* | *4-7* | *8-12* | *13-18* | *19-24* | *Total* |
| N | 1 | 5 | 11 | 3 | | 20 |
| NNE | 5 | 10 | 4 | 1 | | 20 |
| NE | 7 | 9 | 3 | 1 | | 20 |
| ENE | 5 | 6 | 8 | 3 | | 22 |
| E | 2 | 4 | 5 | 1 | | 12 |
| ESE | 3 | 4 | 3 | 1 | | 11 |
| SE | 13 | 8 | 17 | 6 | | 44 |
| SSE | 5 | 19 | 21 | 5 | | 50 |
| S | 15 | 23 | 26 | 6 | | 70 |
| SSW | 6 | 29 | 19 | 2 | | 56 |
| SW | 5 | 44 | 33 | 8 | | 90 |
| WSW | 4 | 17 | 17 | 8 | | 46 |
| W | 8 | 25 | 13 | 8 | | 54 |
| WNW | 4 | 15 | 15 | 14 | | 48 |
| NW | 4 | 3 | 17 | 30 | 6 | 60 |
| NNW | 0 | 6 | 18 | 8 | | 32 |
| Calm | 89 | | | | | 89 |
| | 176 | 227 | 230 | 105 | 6 | 744 |

TABLE 3.4   DEBIASING WIND DATA

| Direction | *Wind Speed Class (MPH)* | | | | | |
|---|---|---|---|---|---|---|
| | *0-3* | *4-7* | *8-12* | *13-18* | *19-24* | *Total* |
| N | 1(1)* | 4 | 9 | 3 | | 18 |
| NNE | 6(5) | 11 | 5 | 1 | | 28 |
| NE | 6(4) | 8 | 3 | 1 | | 22 |
| ENE | 6(4) | 7 | 9 | 4 | | 30 |
| E | 2(1) | 3 | 4 | 1 | | 11 |
| ESE | 3(2) | 5 | 4 | 1 | | 15 |
| SE | 12(5) | 7 | 15 | 5 | | 44 |
| SSE | 6(8) | 21 | 24 | 6 | | 65 |
| S | 13(10) | 21 | 23 | 5 | | 72 |
| SSW | 7(11) | 33 | 22 | 2 | | 75 |
| SW | 4(13) | 40 | 29 | 7 | | 93 |
| WSW | 5(7) | 19 | 20 | 9 | | 60 |
| W | | | | | | |
| WNW | | | | | | |
| NW | | | | | | |
| NNW | | | | | | |
| | 176 | 227 | 231 | 104 | 6 | 744 |

* NOTE:  *Numbers in parentheses are the distributed calms from Table 3.5.*

TABLE 3.5    DISTRIBUTING CALMS

| Direction | $n_w$ (Debiased) | Number of Distributed Calms |
|-----------|------------------|------------------------------|
| N         | 5                | 1                            |
| NNE       | 17               | 5                            |
| NE        | 14               | 4                            |
| ENE       | 13               | 4                            |
| E         | 5                | 1                            |
| ESE       | 8                | 2                            |
| SE        | 19               | 5                            |
| SSE       | 27               | 8                            |
| S         | 34               | 10                           |
| SSW       | 40               | 11                           |
| SW        | 44               | 13                           |
| WSW       | 24               | 7                            |
| W         |                  |                              |
| WNW       |                  |                              |
| NW        |                  |                              |
| NNW       | ___              | ___                          |

concentration in ppb at a parking lot which is located 3,000 feet SSW from the stack. A north wind is blowing at four meters per second and the effective stack height is 150 feet. (Note: for general estimates, it is best to use stability class C or D.)

3.7  Where will the maximum $SO_2$ ground level concentration occur for Problem 3.6 and what will it be in grams per cubic meter?

3.8  If the winds are 1.5 meters per second and fumigation occurs due to 50 meter stable inversion layer, what values of $\sigma_y$ and $\sigma_z$ would be used in Problem 3.6?

3.9  Rework Problem 3.6 using 12 mph winds, a stack height of 125 feet, exit stack gas temperature of 475°F, exit stack gas velocity of 30 feet per second and for a partly clouded nighttime situation. Stack is 10 feet ID.

3.10  You are located downwind from two oil burning power plants. One power plant is located 0.1 mile NNE of your location and burns 2,500 pounds per hour of fuel oil containing 0.5% sulfur. The second power station is located 0.2 mile NNW of you and burns 3,500 pounds per hour of fuel oil containing 0.75% sulfur. Both power plant stacks are 100 feet high. The wind is blowing from the north at a speed of 7 mph. What would the maximum sulfur dioxide concentration be at your location?

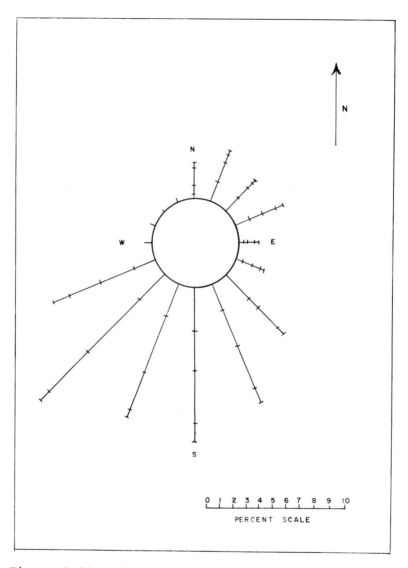

Figure 3.28   Wind Rose--Bias Removed and Calms
              Distributed

3.11   The concentration of hydrogen sulfide is
measured as 50 ppb 100 feet downwind from an
abandoned oil well shaft.   What is the concentration
of the hydrogen sulfide being emitted from the
shaft? (The winds are 5 mph and it is 2:00 on a
sunny June afternoon.)

3.12   Dust particles about 50 microns in diameter
and 3.0 g/cc in density are being deposited on cars
in a parking lot.   A foundry with 100 foot high
stacks is located 500 feet away and is being
suspected of producing the particulate matter.
Another foundry located 1/2 mile away has a 150
foot stack.   If the deposition occurs at night when
there are light 2 meter per second winds, which
plant is producing the pollution that causes most
of the deposits on the cars?

3.13   A privately owned power plant located on the
Ohio River supplies power to the Atomic Energy
Commission.   Assume the following data for this
problem.   Two 30 foot diameter stacks are 150 feet
tall and one 25 foot diameter stack is 250 feet
tall.   Coal consumption ranges from a low of 600
to a high of 800 tons per day.   The coal contains
an average of 2.8% sulfur.   Assume all stacks to
have exit velocities of 10 meters per second at a
temperature of $400 \,^\circ F$.   Calculate for a cloudy fall
afternoon ($45^\circ F$) with a 12 mph wind where the
maximum $SO_2$ ground level concentration will occur.

3.14   What is the maximum concentration of $SO_2$
from Problem 3.13 in ppm?

3.15   A mixture of limestone and coal air pollution
are being released from a stack with an effective
height of 100 meters.   The particles range in size
from 0.5-100 microns.   Where will they strike the
ground on a day with a 4 meter per second wind?
(See Appendix C for density values.)

REFERENCES

1.   Peterson, J.T. "The Climate of Cities, A Survey
     of Recent Literature," United States Department
     of Health, Education and Welfare, Public Health
     Service, National Air Pollution Control Admin-
     istration (NAPCA), Raleigh, North Carolina (1969).
2.   Holzworth, G.C. "Estimates of Mean Maximum Mixing
     Depth in the Contiguous United States," Monthly
     Weather Review, Volume 92, No. 5, pp. 235-242
     (1964).
3.   Hosler, C.R. "Low-Level Inversion Frequency in
     the Contiguous United States," Monthly Weather

Review, Volume 89, No. 9, pp. 319-339 (1961).

4. Sutton, O.G. "A Theory of Eddy Diffusion in the Atmosphere," Proceedings of the Royal Society, Volume 135, pp. 145-165 (1932).

5. Pasquill, F. "The Estimation of Dispersion of Windborne Material," Meteorological Magazine, Volume 90, No. 1063, pp. 33-49 (1961).

6. Gifford, F.A. "Uses of Routine Meteorological Observations for Estimations of Atmospheric Dispersion," Nuclear Safety, Volume 2, No. 4, pp. 47-51 (1961).

7. Thomas, F.W., Carpenter, S.B. and Colbaugh, W.C., "Plume Rise Estimate for Electric Generation Stations," Journal of the Air Pollution Control Association, Volume 20, No. 3, pp. 170-177 (1970).

8. Fay, J.A., et. al., "A Correlation of Field Observations of Plume Rise," Journal of Air Pollution Control Association, Volume 20, No. 6, pp. 391-397 (1970).

# CHAPTER IV
## AIR POLLUTION CHEMISTRY

It is important to be familiar with both
organic and inorganic chemistry in order to more
effectively control air pollution.  Organic
chemistry is especially important because of the
atmospheric reactions of organic compounds.  There
are also inorganic atmospheric interactions, and
as would be expected, inorganic neutralization
reactions are extremely significant for pollution
control.  The emphasis of Chapter IV is placed on
organic and inorganic atmospheric chemical interactions because of the unusualness of these types
of reactions.  It is realized that some practicing
engineers and production personnel do not retain a
high amount of familiarity with organic chemistry
reactions, therefore, the first section can serve
both as a review for those who have forgotten much
of their organic chemistry as well as a basic
organic chemistry introductory session for those
who have had little experience in organic chemistry.
     For those who wish further information in
chemical principles or chemical operations, they
are referred to the following worthwhile publications:  "Riegel's Industrial Chemistry", edited by
J. A. Kent, published by the Reinhold Publishing
Corporation; "Chemistry in Engineering", by L. A.
Murno, published by Prentice-Hall; and "The Condensed
Chemical Dictionary", published by Reinhold.  The
reader who wishes more information in chemical
reaction kinetics is referred to texts as "Chemical
Reaction Engineering" by Octave Levenspiel,
published by Wiley, and "Reaction Kinetics for
Chemical Engineers" by F. M. Wallace, published by
McGraw-Hill.

## 4.1   ORGANIC CHEMISTRY REVIEW

Organic chemistry is concerned with the reaction of compounds that contain carbon atoms combined in various manner with hydrogen, oxygen and/or other atoms or molecules. Carbon compounds can contain atoms of any element, however, hydrogen and oxygen form most of the functional groups in the carbon compounds. Sulfur and nitrogen with oxygen also make other functional groups. Strangely enough, the compounds, carbon monoxide and carbon dioxide, are classified both as organic and inorganic compounds by the Chemical Rubber Company "Handbook of Chemistry and Physics".

Carbon is in Group No. 4 of the Periodic Table of Chemical Elements. As such, it has an electro-negativity of 2.5 which means that it bonds mainly by sharing electrons (covalent bonding) rather than by the stronger ionic type of bonding which gives away or takes on electrons. Organic compounds also enter into coordinate covalent bonding which means that one atom can share all the electrons for both itself and its reaction partner. The remaining two types of bonding, hydrogen bonding and van der Waals forces, are of less significance in the bonding of organic compounds. By being a member of Group 4 (in the middle of the periodic table), carbon can bond with compounds on either the right or left side as well as with itself, giving it the extreme flexibility that it has. Silicon is another member of Group 4 and results in a type of chemistry called silicon chemistry which is similar to carbon (organic) chemistry.

Isomers are compounds that have the same formula (number and type of atoms) but are joined in different ways. This means that they are completely different compounds although they do have the same number of atoms of various elements in their structure. For example, n-butane and iso-butane have the same molecular formula of $C_4H_{10}$. However, n-butane is a straight-chain compound with the structural formula:

$$CH_3-CH_2-CH_2-CH_3$$

and iso-butane is a branched chain compound with the structural formula:

$$CH_3 - \underset{\underset{CH_3}{|}}{\overset{\overset{H}{|}}{C}} - CH_3$$

The boiling point of n-butane is -0.5°C. The boiling point of iso-butane is -11.7°C.

Another example is the molecule $C_2H_6O$, which can be either dimethyl ether, which has a structural formula of:

$$CH_3 - O - CH_3$$

or ethanol which has a structural formula:

$$CH_3 - CH_2 - OH$$

The more atoms in a particular organic molecule, the greater the possible number of isomers that can exist. For instance, the formula $C_9H_{20}$ has thirty-five known isomers.

Hydrocarbons are a special class of organic compounds which contain only carbon and hydrogen. They are relatively inert (even though they enter into many atmospheric reactions) and many isomers of the hydrocarbons exist. Hydrocarbons have two general classifications: aliphatic--which are unclosed chain type compounds (also known loosely as the paraffin series), and aromatic--which are closed or ring type structured compounds. Both classes contain paraffin or paraffinic series which may be either saturated or non-saturated.

4.1.1 ALIPHATIC HYDROCARBONS (PARAFFIN SERIES)

*Paraffins or Alkanes*

These are *saturated* paraffin hydrocarbons, which means that they contain as much hydrogen as possible (all other aliphatic hydrocarbons are *unsaturated* paraffins). The simplest member of this series is $CH_4$, which is methane. The general formula of this series is $C_nH_{2n+2}$.

A special variation of this group is the cyclo-alkanes of which cyclohexane is a typical example. The formula for cyclohexane is:

$$
\begin{array}{ccc}
 & CH_2 & \\
H_2C & & CH_2 \\
H_2C & & CH_2 \\
 & CH_2 & \\
\end{array}
$$

The general formula for the cyclo-alkanes is $C_nH_{2n}$. Densities and boiling points of the cyclic paraffins (also called alicyclics) are higher than for the corresponding open chain compounds.  These compounds and their derivatives exist naturally in crude oils and are called naphthenes (not naphthalene which is mentioned in Section 4.1.2).

These are the only saturated aliphatic compounds. From here on, the groups are unsaturated, which means they do not contain as much hydrogen as they could.

*Olefins or Alkenes*

The simplest member of this group is ethylene or ethene and has the structural formula:

$$CH_2 = CH_2$$

The general formula for this group is $C_nH_{2n}$.  Special variations of this group include diolefins, which have two double bonds in the compound structure (and, therefore, there are two less hydrogen in the general formula) and triolefins, which have three double bonds in the compound (and, therefore, the general formula contains four less hydrogens).

*Acetylenes or Alkynes*

Acetylene or ethyne is the simplest member of this group and has the structural formula:

$$CH \equiv CH$$

The general formula for this class is $C_nH_{2n-2}$.  It should be pointed out that the more unsaturated the compound, the more reactive it is.  Unsaturated compounds revert to saturated conditions as rapidly as possible and in doing so can enter into atmospheric reactions.

4.1.2  AROMATIC HYDROCARBONS

Aromatic organic compounds are cyclic *un*saturated hydrocarbons.  The name of this group was obtained from the fact that many of these compounds have strong aromas.  This name, in itself, warns of a potential air pollution problem from these compounds. A typical example of this group is benzene, which has a formula $C_6H_6$, and can be represented structurally by:

CH
HC — CH
HC — CH
CH

Polycyclic organic compounds have more than one ring to the structure. This type of organic chemistry is also nicknamed "chickenwire chemistry" because of the obvious appearance of the compound formulas when expressed in a two-dimensional structural form. Two common polycyclic compounds are naphthalene, which appears as two joined benzene rings:

and anthracene, which appears as three benzene rings joined:

## 4.1.3 FUNCTIONAL GROUPS WITH OXYGEN

Addition of oxygen to hydrocarbons gives a vast arrangement of new types of compounds.

*Alcohols*

Alcohols contain the functional hydroxyl group (-OH). Boiling points of alcohols are higher than those of hydrocarbons indicating that alcohols should be less likely to become atmospheric pollutants. The simplest alcohol is methanol, which has the structural formula:

$$CH_3OH$$

The general formula for the simple alcohols is ROH, where R is an alkyl radical. A special group of alcohols called glycols have two -OH functional groups. Ethylene glycol (antifreeze) is an example of a glycol and has a structural formula:

$$
\begin{array}{ccc}
CH_2 & - & CH_2 \\
| & & | \\
OH & & OH
\end{array}
$$

Another special type of alcohols are the glycerols. These compounds have three -OH functional groups. Glycerine is an example:

$$CH_2 - CH - CH_2$$
$$\;|\qquad\;|\qquad\;|$$
$$OH\quad OH\quad OH$$

The glycols and glycerols are thicker than the alcohols and are, therefore, not as volatile. For this reason, they contribute even less to air pollution problems than the alcohols.

*Aldehydes*

Aldehydes have the functional group - $C\overset{\nwarrow O}{-}H$. Aldehydes are extremely reactive compounds and as such, create many air pollution problems. Formaldehyde is the simplest member of this group and has the structural formula:

$$H - C\overset{\nwarrow O}{-} H$$

Acetaldehyde, which is frequently present in photochemical smog is:

$$CH_3 - C\overset{\nwarrow O}{-} H$$

Aldehydes are very reactive and are easily oxidized to organic acids.

*Ketones*

Ketones have the general formula:

$$R - C\overset{\nwarrow O}{-} R'$$

(The meaning of R- and R' - is explained in Section 4.1.4.) Acetone is an example of a simple ketone and has the structural formula:

$$H_3C - C\overset{\nwarrow O}{-} CH_3$$

Ketones are volatile and flammable and create air pollution problems. Ketones are also slightly polar and, therefore, have solubility in both water and organic liquids.

*Organic Acids*

Organic acids have the function group:

$$-C\overset{\nwarrow O}{-} OH \text{ (or written - COOH)}$$

The - COOH group is also known as a carboxylic radical. An example of an aliphatic acid is formic acid which has the formula $HCOOH$; and an example of an aromatic acid is benzoic acid, which has the formula $C_6H_5COOH$. Long chain aliphatic acids, such as oleic and stearic acids which are obtained from animals or plants, are called fatty acids. These compounds are well known for pollution problems. Salts of these acids are called soaps and soap factories are known for odor problems. Aromatic acids such as phenol (carbolic acid) are odorous air pollutants. Phenol has the formula $C_6H_5OH$ where the -OH group is attached to a hydrogen in the aromatic ring in this case.

*Ethers*

Ethers have the general formula ROR. They are highly volatile and dangerous pollutants. The simplest ether is dimethyl ether which has the formula:

$$CH_3 - O - CH_3$$

The compound we know as ether is really ethyl (or diethyl) ether which has the formula:

$$(C_2H_5)_2 O$$

*Esters*

Esters have the general formula:

$$R - C\overset{\nwarrow O}{-} O - R'$$

and can be formed by the reaction of a carboxylic acid and an aldehyde. A simple ester is ethyl acetate which has the formula:

$$CH_3 - C\overset{\nearrow O}{-} O - C_2H_5$$

Esters are used for solvents, dopes and drying oils and are, therefore, obviously air pollutants.

## 4.1.4  RADICALS

Aliphatic (paraffin series) hydrocarbon radicals are called alkyl radicals and are usually represented by "R-" or "R'-" which may or may not be the same. They are obtained by removing one hydrogen from the formula. The ethyl radical is $C_2H_5-$; the n-propyl radical is $CH_3-CH_2-CH_2-$. Aromatic radicals, known as aryl radicals, are obtained the same way.

Acyl radicals are obtained by replacing the -OH group in an organic acid by some other constituent. Note that this amounts to the same as removing a hydrogen from an aldehyde.

## 4.1.5  OTHER FUNCTIONAL GROUPS

Elements such as nitrogen and sulfur combined with oxygen and/or hydrogen are functional groups. The most important of these as far as air pollution control is concerned are the groups formed with nitrogen.

*Amines*

Organic amines have the general structural formula:

$$RNH_2$$

and are, therefore, derivatives of ammonia ($NH_3$). When only one hydrogen in the ammonia molecule is replaced by an organic radical, it is a primary amine and appears as indicated in the structural formula just given. When two of the hydrogens are replaced, it is a secondary amine, and when three of the hydrogens are replaced, it is a tertiary amine. An example of a primary amine is aniline with the structural formula:

$$C_6H_5NH_2$$

Aniline is a highly odorous compound.

*Nitrates*

Organic nitrates have the general structural formula:

$$RNO_2$$

and are extremely detrimental air pollutants. The complex organic nitrate known as PAN (peroxy acetyl nitrate):

$$CH_3 - C \overset{\displaystyle \nearrow^O}{} - O - O - NO_2$$

is formed by atmospheric reaction of nitrogen oxides and organic comppounds in the presence of sunlight. PAN also represents the group of compounds known as peroxy acyl nitrates, indicating that they are formed from organic acids as well as from acetalde-hyde as indicated here.

*Mercaptans*

Mercaptans are an example of organic compounds containing sulfur as part of the functional group. The structural formula of mercaptans resembles alcohols except that the oxygen is replaced by sulfur. An example is ethyl mercaptan, $C_2H_5SH$ which has an extremely strong and disagreeable skunk-like odor. Mercaptans are important in air pollution because of their odors and not because of their ability to react.

## 4.2  NITROGEN OXIDES

Nitrogen has oxidation numbers ranging from -3 to +5. This makes it possible for nitrogen to form many different oxides as well as other chemical compounds. Nitrogen oxides ($NO_x$) consist mainly of NO and $NO_2$. Until recently, literature reported that the usual ratio of $NO_2$ to NO was 2:1. Recent data indicate the NO concentration dominates at night and the concentration of NO and $NO_2$ is approximately equal in the atmosphere during the day.

Nitrogen oxides are formed by the oxidation of nitrogen or nitrogen compounds that contain nitrogen in a low oxidation state. Nitrogen gas ($N_2$) in the atmosphere has an oxidation number of zero. When air containing nitrogen is heated in the presence of oxygen, it can react to form a nitrogen oxide. Ammonia ($NH_3$) contains nitrogen in an oxidation state of -3. Ammonia can also be reacted with air or oxygen to form nitrogen oxides. In addition, ammonia, which is a colorless, odorous gas, is in itself a very irritating poisonous air pollutant. Nitrogen is also contained in other compounds and even exhibits apparent fractional average oxidation

numbers.  Most of the nitrogen oxides in the
atmosphere result from high temperature combustion
of air in internal combustion engines or in fossil
fuel burning power generating stations (see Section
8.2).  Table 4.1 lists the oxides of nitrogen and
shows some of the properties exhibited by these
compounds at atmospheric conditions.

TABLE 4.1   OXIDES OF NITROGEN

| *Formula* | *Name* | *Oxidation No. of N* | *Properties at Atmospheric Conditions* |
|---|---|---|---|
| $NO\times$ | nitrogen oxides | $\times$ | essentially NO + $NO_2$; not including $N_2O$. |
| $N_2O$ | nitrous oxide (laughing gas) | +1 | colorless, odorless gas; toxic; relatively stable; normal atmospheric concentration = 0.5 ppm. |
| NO | nitric oxide | +2 | colorless, odorless gas; highly toxic; not too stable; product of high temperature combustion. |
| $N_2O_3$ | nitrogen trioxide | +3 | slightly bluish gas; very toxic; forms $HNO_2$ with water; dissociates to $NO+NO_2$. |
| $N_2O_4 \rightleftharpoons NO_2$ | nitrogen dioxide | +4 | reddish-brown gas; pungent odor; toxic; formed by reaction of NO with $O_2$ at 20 to $160\,^{\circ}C$; forms $HNO_3$ with water. |
| $N_2O_5$ | nitrogen pentoxide | +5 | white crystalline solid; unstable; forms $HNO_3$ with water. |

## 4.3 ATMOSPHERIC REACTIONS

### 4.3.1 WITH NITROGEN OXIDES

The atmosphere contains oxidants which are chemicals capable of entering into oxidation-reduction type reactions and thereby oxidizing elements or compounds. Ozone ($O_3$) is an oxidant and exists naturally in the atmosphere. Ozone, as well as most other atmospheric oxidants in polluted air, are formed by chemical reactions which occur among the primary pollutants. Sunlight acts as a catalyst to increase the rate of some atmospheric chemical reactions. The sunlight catalyzed reactions are called photochemical reactions. Smog, which originally was thought to be a combination of smoke and fog, can be produced by photochemical reactions and is made up of numerous chemicals including oxidants. In instances where separate measures of ozone and total oxidant are available for the same sample of photochemical smog, the data indicate that *most of the oxidant* in smog *is ozone*. The presence of this ozone is evidenced by the cracking of stressed rubber.

A California chemist, Haagen-Smit (Director of the Los Angeles Air Pollution Control Agency), discovered that the nitrogen oxides and hydrocarbons present in the Los Angeles atmosphere (from auto exhaust) reacted in the presence of sunlight with oxygen to form ozone. He demonstrated this by irradiating this type of atmosphere with artificial sunlight.

The main constituents that react in oxidizing smog are $O_3$, NOx, organic acids, aldehydes, and unsaturated hydrocarbons of both olefinic and aromatic types. Ozone is both a reaction product and an intermediate in the steady state series of atmospheric reactions which occur as follows:

1. The relatively unstable NO reacts with oxygen:

$$2NO + O_2 \xrightarrow[20-160^{\circ}C]{} 2NO_2$$

This reaction can occur during the night as well as during the day and is what accounts for the creation of a brown-colored haze present in the atmosphere over areas having photochemical smog.

2. In the sunlight catalyzed decomposition reaction, the nitrogen dioxide formed breaks down:

$$NO_2 \xrightarrow[\lambda < 4200 \ A^O]{h \ \nu} NO + O$$

3.   The very reactive oxygen atoms formed from this reaction react extremely quickly with molecular oxygen in the atmosphere to form ozone:

$$O + O_2 + M \longrightarrow O_3 + M$$

M is any other molecule present in the atmosphere and is required to absorb the energy of the reaction. It is necessary for three molecules to strike together at the same time for this reaction to take place.   Even though this is difficult to accomplish, the oxygen atoms react rapidly and have a half-life of only ten millionths of a second in the air (1).
4.   The ozone and NO produced in the last two reactions of this reaction chain can react together to produce $NO_2$ and $O_2$:

$$NO + O_3 \longrightarrow NO_2 + O_2$$

5.   Hydrocarbons present in the atmosphere prevent a continuous recycle and permit ozone to accumulate by reacting with some of the nitric oxide which would otherwise react with the ozone.   The process by which the hydrocarbons remove the nitric oxide is to first react with ozone according to:

$$O_3 + HC \longrightarrow RC \overset{\displaystyle\nearrow O}{-} H + RC \overset{\displaystyle\nearrow O}{-} O\cdot$$

(The symbol R in these formulas represents an alkyl radical.)   The products of this reaction are aldehydes and oxyacyl radicals.
6.   The singlet active oxygen atom from Step 3 also reacts with hydrocarbons giving:

$$2O + HC \longrightarrow R\cdot + RC \overset{\displaystyle\nearrow O}{-} OH$$

The products of this reaction are alkyl radicals and organic acids.   Acids and the aldehydes, which are present in photochemical smog can be produced by reactions such as the two previous equations or they can be released directly into the atmosphere as primary pollutants.
7.   The activated radicals in turn can react with ozone:

$$R\cdot + O_3 \longrightarrow R\overset{\diagup O}{\diagdown} - O - O\cdot$$

This product is a peroxy acyl radical. Any combination of Steps 5, 6, & 7 must produce *two* radicals or else none are formed.

8. Both the oxy acyl radicals from Step 5 and the peroxy acyl radicals from Step 7 react with some of the nitric oxide and thereby prevent it from reacting according to Step 4 and removing ozone. These reactions are:

$$RC\overset{\diagup O}{\diagdown} - O\cdot + NO \longrightarrow RC\overset{\diagup O}{\diagdown} + NO_2$$

$$R\overset{\diagup O}{\diagdown} - O-O\cdot + NO \longrightarrow R\overset{\diagup O}{\diagdown} - O\cdot + NO_2$$

The first reaction forms an acyl radical and the second reaction forms an oxy acyl radical, both producing $NO_2$.

9. The final step in this chain of reactions that originates because of photochemically catalyzed reactions producing NO, O, and ultimately $O_3$ in the atmosphere gives:

$$RC\overset{\diagup O}{\diagdown} - O-O\cdot + NO_2 \longrightarrow RC\overset{\diagup O}{\diagdown} - O - O - NO_2$$

The products of this reaction are peroxy acyl nitrates (PAN). If the alkyl radical R- represents the simplest alkyl radical $CH_3-$, then this final compound would be peroxy acetyl nitrate.

Stephens (2) attempts to summarize these reactions with a chart showing the steady state relationship between ozone, nitric oxide, and nitrogen dioxide with the influence of the accompanying *slow* side reactions given above as Steps 5 through 9. It should be noted that PAN is also an atmospheric oxidant but concentrations are low compared to $O_3$.

## 4.3.2 WITH SULFUR DIOXIDE

Wilson (3) reports that atmospheric reactions with $SO_2$ vary depending upon whether the photochemical smog is present. In the absence of smog, the reactions of $SO_2$ depend primarily on the amount of moisture present. In dry air, there is no

apparent loss of $SO_2$, either with or without sunlight energy. The $SO_2$ becomes activated (energy
level of the molecule is increased), but it does
not appear to dissociate or react:

$$SO_2 \xrightarrow[\lambda > 3000A^\circ]{h\nu} SO_2{*}$$

At relative humidities greater than about 30%, the
activated $SO_2$ is oxidized to $SO_3$:

$$SO_2{*} + \tfrac{1}{2} O_2 \longrightarrow SO_3$$

The $SO_3$ reacts with water vapor to form sulfuric
acid mist:

$$SO_3 + H_2O \longrightarrow H_2SO_4$$

This acid mist in the atmosphere is considered to
be more toxic to humans than $SO_2$. The rate of
production of the acid increases with both
increased humidity or $SO_2$ concentration.

In the presence of photochemical smog, the $SO_2$
reacts with $NO_2$ and/or the ozone oxidant present
and water vapor to form sulfuric acid mist:

$$SO_2 + NO_2 \longrightarrow SO_3 + NO$$

$$SO_2 + O_3 \longrightarrow SO_3 + O_2$$

$$SO_3 + H_2O \longrightarrow H_2SO_4$$

The rate of reaction of the $SO_2$, as measured by the
disappearance of $SO_2$, is much faster when $O_3$ is
present. This reaction, which removes $O_3$, also
serves to slow down the rate of photochemical smog
reactions (given in Section 4.3.1) because it
reduces the existing oxidant ($O_3$) concentration.

4.3.3 REACTION RATES

Reaction rates in the atmosphere with the
photochemically produced ozone, have approximately
second-order reaction rates. The rate of reaction
of a given component in ppm/min is:

$$r_a = -\frac{dC_a}{dt} = k\, C_a^{\,2}$$

TABLE 4.2  REACTION RATES WITH $O_3$ (4)

| Pollutant | Organic Type | Half Life in Min. at O3 Conc: 0.2 ppm | 1 ppm | $k$ (at 25°C), $ppm^{-1} min^{-1}$ |
|---|---|---|---|---|
| ethylene | olefin | 1,100 | 220 | 0.0045 |
| cyclohexane | cyclo olefin | 57 | 12 | 0.087 |
| typical gasoline | paraffin-olefin-aromatic mix | 380 | 76 | 0.013 |
| acetylene | acetylene | 500,000 | 24,000 | 0.0001 |
| NO | / | 0.16 | 0.03 | 32.0 |
| $NO_2$ | / | 65 | 13 | 0.077 |

where: $C_a$ = concentration of initial component
a, ppm
t = time, min.
k = specific reaction rate, $ppm^{-1} min^{-1}$

Table 4.2 lists the specific reaction rates of
various atmospheric pollutants with ozone. A
maximum ozone concentration of about 1 ppm is the
limit to the amount of ozone which can be formed
in ambient polluted air.

SUGGESTED ADDITIONAL READING

The reader is referred to the following two
texts for supplemental information on air pollution
chemistry: Stern's (1) Volume 1 "Air Pollution"
and "The Air Pollution Handbook" by Magill (5).

QUESTIONS FOR DISCUSSION

1. Describe a specific photochemical reaction and
show why it is important in air pollution studies.
2. Plot schematically the chain of reactions
producing PAN from nitrogen oxides in the atmosphere.
Connect related reactions as follows: use a solid
line for the fast reactions and dotted lines for
the related slow side reactions.

PROBLEMS

4.1 A typical metropolitan atmosphere may contain
a total of 2 ppm of organic pollutants. Condensable
organic pollutants were obtained and data were
reported as follows: cond. organic pollution of
the air equals 0.5 ppm--55% paraffins, 4% cyclo-
paraffins, 14% olefins, 7% acetylenes, 5% aromatic
hydrocarbons, 2% chlorinated hydrocarbons, balance
other compounds.
    a) List by groups what specific organic
compounds may be indicated by each type listed above.
    b) Using the whole percentage values given
above, what percent by volume of the air is composed
of these organic pollutants (remember mole percent
equals volume percent for ideal gases).
    c) If the air contained 0.5 ppm $NO_x$, show what
might be expected to happen.

REFERENCES

1. Stern, A. C., "Air Pollution," Vol. 1, Second
   Edition, Academic Press (1968)

2.  Stephens, E. R., "Chemistry of Atmospheric Oxidants," Journal of Air Pollution Control Association, Vol. 19, No. 3, pp. 181-185 (1969)
3.  Wilson, W. E., Jr., and Levy, A., "A Study of Sulfur Dioxide in Photochemical Smog," Journal of Air Pollution Control Association, Vol. 20, No. 6, pp. 385-390 (1970)
4.  Katz, M., "Air Pollution" (Monograph Series of the World Health Organization), Columbia University Press, Chap. 5, p. 152 (1961)
5.  Magill, P.L., Holden, F. R., Ackley, C., "Air Pollution Handbook," McGraw-Hill (1956)

# CHAPTER V

# EFFECTS

Air pollution causes many effects, some of
which are immediate and obvious (sore eyes,
difficult breathing, neucrotic vegetation), and
nearly all of which could result in chronic
degradation of man and his resources under specific
conditions. The net results of pollution on man
can be said to be economic and/or time effects.
The economic effect is the direct and indirect
costs of controlling pollution. The immediate cost
of providing pollution controls is an economical
burden, but pollution control costs can be received
back, sometimes with a profit return. The time
consideration implies length of life. Most data
show that air pollution is detrimental to life,
but it is also known that some pollutants are
beneficial.

The three general areas into which air
pollution effects can be classified are: 1)
effects on vegetation, 2) physiological effects
on man and other animals, and 3) material effects.
Excessive pollution can cause plants and crops to
be bleached, discolored and stunted--small amounts
of pollution can increase growth rate. Estimates
of costs due to excessive air pollution damage in
the United States vary from $4 to $20 billion per
year. No income costs are made for benefication
of plants by pollution or for secondary effects
of poisoning of animals who eat poisoned plants.

Superficial physiological damage to man and
other animals occurs in the form of irritation of
the eyes, skin, upper respiratory tract, stomach
(nausea) and sinuses (allergy). Air pollutants
as a whole can literally affect every part of the
body tissues and organs promoting emphysema and
overburdening the heart. People who are

incapacitated because of air pollution (and
smoking) inflicted illnesses, such as the incurable
emphysema, may not be able to work and earn wages.
In 1965, it was reported that the State of California had over 670,000 chronic respiratory disabili-
ties, including 25,000 permanently disabled,
because of air pollution. That year, it was
estimated that 3,000 people died in California
from air pollution. Emphysema is the fastest
rising cause of death in the United States. Small
amounts of certain pollutants can be beneficial.
For example, $SO_2$ is a disinfectant and destroys
harmful bacteria.

It is possible, but difficult, to show that
air pollution costs money because it destroys
property: it does cost money to clean and launder
pollution soiled clothing but clothing should be
washed periodically anyway. It costs money to
clean or paint buildings and to repair structures
that are damaged by corrosion, erosion, oxidation
or spattering resulting from air pollution. It is
estimated that air pollution damages cost every
person in the United States $65 a year, but no one
really knows what the true pollution damage costs
are.

## 5.1   EFFECTS ON VEGETATION

Plant damage, as a result of exposure to air
pollutants, varies with the plant and the type of
pollutant as well as with the time of day of
exposure and the concentration of pollution in the
atmosphere. Threshold concentrations which cause
damage to plants vary with the length of exposure
of the plant to the pollutant. Also, there is a
time lag which exists from the time of exposure
until the time when the symptoms show up. This
sometimes makes it difficult to relate the pollution
damage to a field plant with a specific pollutant
and pollution exposure period. Plant insects and
diseases can cause damage which appears similar to
certain types of pollution injury.

The mechanism by which pollutants are phytoxic
(poisonous to plants) results mainly from altered
plant enzyme activity. It appears that the enzyme
activity can be inhibited and then increased. This
damages the plant by first lowering the respiration
then increasing the respiration. The enzyme enolase
appears to be effected most by the pollutants.
This enzyme controls the reaction:

$$
\begin{array}{c}
\underset{\big\backslash}{COO^-} \\
HCOPO_3^- \\
\big| \\
H_2COH
\end{array}
\quad\xrightarrow[\text{Catalyst}]{\text{Enolase}}\quad
\begin{array}{c}
\underset{\big|}{COO^-} \\
COPO_3^- + H_2O \\
\big\| \\
CH_2
\end{array}
$$

Plants are used to indicate the presence of air pollutants. Plants near heavily traveled highways and down wind from polluted areas are definitely influenced by the pollution. Some industrial operations have been known to set out susceptible plants as indicators to show the public that they are controlling their pollution.

## 5.1.1  LEAF STRUCTURES

Leaves have two major functions: to feed the plant and to protect itself. Green leaves contain chlorophyll which, with the help of sunlight, changes carbon dioxide and dissolved chemicals into sugar compounds, oxygen and water. Liquids travel throughout the leaf by means of the veins. Veins also connect the leaf to the stem as well as give the leaf structural support. The narrow flexible stems allow the leaf to flutter without snapping off in high winds. This fluttering speeds up evaporation of the water produced. The external structure of a leaf usually also contains drain tips. These tips are natural rain spouts that permit drainage of excessive rain water that could otherwise damage a leaf. Species from arid regions do not usually have these drain tips because they must conserve the water.

Figure 5.1 is a cross-sectional view of a portion of a leaf. The lower epidermis is a single layer of cells which contain no chlorophyll except for the guard cells. Guard cells surround the openings in the leaf which are called stomata or stomates. The stomates are the avenues of exchange permitting gases to enter and leave the leaf. There are 100 to 350 stomata per square millimeter of lower leaf surface. The guard cells have the ability to cause the stomata to open or close as the light intensity decreases or increases. The top surface of the leaf contains cutin which is a transparent waxy coating. The sunlight easily passes through the cutin. Cutin reduces evaporation from the top surface of the leaf and also protects the leaf against insects and abrasive pollutants which settle on the leaf.

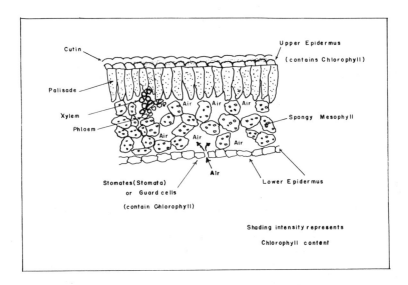

**Fig. 5.1   Internal Cross Section of a Foliage Leaf**

Beneath the cutin is the upper epidermis which is
a single layer of cells that contain chlorophyll.
Leaves also contain xylem and phloem--the xylem
conduct water and mineral salts upward into the
mesophyll and the phloem conduct foods downward to
the petiole which is a manifold of vascular
bundles which connect to the veins.  The mesophyll
is a spongy tissue also rich in chlorophyll and,
with the palisade, is the food making tissue of
the plant.  It should be remembered that both
stems and bark (cambrium layer) also have stomata.

## 5.1.2  MODES OF POLLUTION ATTACK

Gases usually enter plants through the stomata
on the leaves or stems, however, they can become
dissolved in water on the leaves or enter through
the roots from the soil.  Gases are usually
absorbed into the mesophyll area of the leaves.
Different species of plants vary considerably in
their susceptibility to injury by gaseous pollutants, the differences being due to the different
rates of absorption of the gas by the leaves.
Environmental factors influence when a particular

plant will absorb gases. Factors which cause
stomata to open more are adequate ground moisture
and nutrients, high humidity, moderate temperatures,
moderate sun intensity and increased $CO_2$ concentration. The stomata close when there  excessive
sun intensity, high temperatures, dry conditions
and darkness. It has been determined that light
intensity differences between field and greenhouse
conditions have no apparent harmful or beneficial
effect on plants as far as their susceptibility to
air pollutants. Gaseous pollutant effects on
plants vary depending on pollutant type, pollutant
concentration, length of exposure and time of day
of exposure as well as other environmental factors.
Hydroponic plants (growing with roots exposed to
air) can take up pollutants, such as ozone,
directly from the air through the roots.

Particulate type air pollutants such as ash,
dirt and grit, land on the top of leaves. They do
not enter the leaf but may damage it by mechanical
abrasion of the surface of the leaf. Particulate
matter can also block out sunlight and thereby
reduce the foodmaking ability of the plant. Up to
100 tons per square mile per month of particulates
land in certain areas. Areas on the leaf surface
damaged by particulate abrasion are more susceptible
to attack by chewing bugs.

Air pollution damage to plants resulting from
the pollution in a particular urban area is not
confined to the immediate locale. Effects on
trees, for instance, in California are noticed
over eighty miles from the metropolitan region.
Even in "remote" areas, such as Canada, air
pollution damage to plants is reported thirty to
forty miles downwind from urban Toronto.

## 5.1.3  SPECIFIC POLLUTANTS

*SULFUR DIOXIDE*

Sulfur dioxide is a major plant damaging agent.
Small amounts of sulfur in a plant are necessary for
plant life, however, large concentrations cause
exosmosis or plasmolyses (meaning an excessive
amount of water flows from the plant). Sulfur
dioxide is absorbed into the mesophyll area of the
plant after entering thru the stomata. Plants
convert sulfur dioxide to the sulfite ion ($SO_3^=$)
then to the sulfate ion ($SO_4^=$)

$$SO_2 \longrightarrow SO_3^= \longrightarrow SO_4^=$$

Large concentrations of sulfur dioxide entering a plant cause a high sulfite ion concentration to exist because in the plant the $SO_2$ can be converted to the sulfite ion more rapidly than the sulfite ion can be converted to the sulfate ion. Plant damage results from the high $SO_3^=$ concentration because $SO_3^=$ is thirty times more toxic to plants than $SO_4^=$. Therefore, the toxicity is due to the reducing ability of $SO_2$ not the acidity.

Acute damage due to $SO_2$ is localized, and the injured areas on the leaf never recover. Uninjured areas of the plant continue operation and new leaves that grow after the $SO_2$ damage occurs are not effected. Acute injury appears as brownish-yellow spots on the surface of the leaf. Acute damage occurs approximately at $SO_2$ conc. 0.03 ppm; chronic injury occurs at lower concentrations. Chronic damage appears as brownish-red or white spots on the surface of the leaf. Plants most sensitive to $SO_2$ damage are alfalfa, barley, cotton, wheat and apples. The most $SO_2$ resistant plants are potatoes, onion, corn and maple trees.

Pollution damage to trees often remains permanently obvious. Tree growth, as evidenced by the internode distance, is shown in Figure 5.2. $SO_2$, for example, slows tree growth. Power plants using fossil fuel cause trees located in the dominant downwind direction to grow more slowly. It is sometimes possible to tell how many years the power plant started operation by visually inspecting trees for years of retarded growth.

*FLUORIDE*

Fluoride damage occurs at very low concentrations (0.1 ppb). Fluoride is a cumulative poison and, therefore, unlike $SO_2$. The fluoride ions that are absorbed into the leaf mesophyll are transported to the leaf by diffusion, and concentrate there. Edges of leaves often have 50 to 200 ppm fluoride concentrations. (Animals that eat these plants are likely to get fluorosis which causes mottling of teeth and softening of bones.) Flouride damage appears as brownish-yellow tips or edges on the leaves.

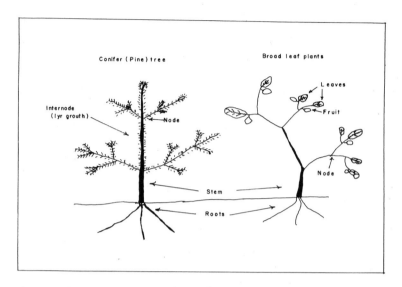

Fig. 5.2. Tree Growth Nodes

Fluorine, like chlorine, is a halogen. However, the fluoride ion is unlike the chloride ion in that it cannot be picked up from the soil (chloride ions can be picked up by the plant from the soil). Fluoride pollution can come from glass industries, aluminum smelters, super-phosphate manufacturers and even coal burning.

The most sensitive plants to fluoride are gladiolus (dark and light), apricot, prune and pine. The most resistant plants are linden, elm, pear, apple, alfalfa and rose. Certain aluminum smelting operations have been known to plant gladioli on the plant grounds to show the public that as long as the gladioli are healthy, they are controlling their fluoride emissions.

*OZONE*

Ozone enters the stomates and damages the palisade area of leaves. Ozone even enters plants during the nighttime when the stomates are supposedly closed (potatoes are extremely susceptible to ozone damage because their stomates are open at night). Ozone causes the palisade cell

walls and the epidermis to collapse, causing
neucrosis, or dead appearing plants.  Ozone causes
pine needles to have neucrotic tips (similar to
$SO_2$ damage but with no distinct line between the
green and the dead areas).  Ozone oxidizes glucose,
therefore, plants that have more sugar are more
resistant to ozone damage.  Damage occurs to some
plants at concentrations less than 0.02 ppm.

Ozone damage appears as a stipple on the
upper surface of the leaf.  The many small stipple
dots appear as either reddish-brown or bleached
white flecks resulting from the collapse of the
palisade.  Insect chewing also causes leaf flecks,
but they are more uniform and spread all over the
leaf.  The most ozone sensitive plants are
tobacco, tomato, bean, spinach, potato and oats.
Ozone resistant plants include mint, geranium,
gladiolus and pepper.

*PAN*

PAN (peroxyacetal or peroxyacyl nitrates)
enter the leaf through the stomata and causes the
mesophyll to shrink, dehydrate, and then fill with
air.  The partially collapsed leaf damage may appear
to extend throughout the leaf thickness, however,
most of the damage is visible on the underside of
the leaves.  Grasses damaged by PAN may appear
more chlorotic (yellow) than neucrotic (dead).
PAN does not usually damage plants if they are not
first exposed to light.  PAN damage usually occurs
at concentrations greater than 0.01 ppm.

PAN damage appears as extensive glazed, silver
or bronze-colored areas on the underside of
the leaf.  The most PAN sensitive plants are
petunia, lettuce, pinto bean and bluegrass.  PAN
resistant plants include cabbage, corn, wheat and
pansy.

*MISCELLANEOUS*

*Nitrogen dioxide* damages plants in a manner
similar to that of sulfur dioxide.  Nitrogen
dioxide is known to retard plant growth at concentrations greater than 0.5 ppm.  Acute damage usually
occurs at concentrations less than 1.0 ppm.

*Hydrogen sulfide* ($H_2S$) injures the growing
tips of plants that are exposed to an acute
concentration of this pollutant although resistant

plants show no damage at 400 ppm for 5 hours (1).
This is one pollutant for which man is *much* more
susceptible than are plants.

*Chlorine* bleaches the leaves of vegetation.
Chlorine damage may look similar to that of either
$SO_2$ or ozone, although chlorine is three times more
phytoxic than $SO_2$. Chlorine damages some plants in
two hours' time at concentrations of 0.1 ppm (1).
Chlorine damage is worse than *hydrochloric acid
(HCl)* vapor damage to plants. (HCl damage looks
like PAN leaf damage.)

*Ammonia (NH₃)* and *hydrochloric acid (HCl)*
cause "acid burn on plants at high concentrations;
2.5 ppm HCl causes a detectable respiration decrease
in plants. The damage from these pollutants is
similar to that of $SO_2$ (not cumulative). These
pollutants arrest photosynthesis and cause
bleached or scorched appearing leaves. As mentioned
earlier, the chloride ion can be picked up by the
plant from the soil.

*Hydrocarbon* gases (e.g. ethylene, acetylene
and propylene) are also known to damage plants,
however, data on this is scarce. It is known that
ethylene has damaged orchids and carnations in
greenhouses. The damage appears as scorched tips
on the flowers and occurred at concentrations greater
than 5 ppm. *CO* affects plants similarly to
ethylene but only when the concentration of CO is
greater than 500 ppm.

## 5.2  EFFECTS ON MAN AND OTHER ANIMALS

Air pollution affects man internally when he
inhales and ingests it as well as externally when
it comes in contact with the body. In addition to
the actual direct contact, animals ingesting
polluted vegetation can pass the pollution on to
man. For example, cows that are permitted to
graze on grass containing fluoride, lead or
radioactive type pollutants can pass them on to
anyone who drinks milk or eats meat.

The use of man as an experimental animal to
determine the effect of pollution is limited to
specific instances. Data of pollution effects on
man can be obtained by epidemiological studies,
field studies, clinical studies or by direct
experimental studies. The epidemiological approach
provides us with data resulting from air pollution
disasters (for example, London, Denora, Los Angeles).
During this type of study, it is necessary to

examine the work habits, social background,
nutritional conditions and all other variables
before it is possible to begin to interpret the
results.

Field studies provide information gathered
from people who are engaged in their normal activities. This type of study makes it possible to
talk to the "man-on-the-street" and ask him
questions and record his results. Field studies
include submitting questionnaires to coaches
requesting information on athletic performance
during certain periods when there are high levels
of air pollution as compared with normal levels of
pollution. In this type of study variables such
as environmental factors of temperature, pressure,
humidity and solar intensity must be considered
as well as mental attitude of the subjects. Another
source of valuable field data is industrial medical
records.

The problem of air pollution effects on man
are sometimes examined using clinical studies. In
this type of study, it is necessary to differen-
tiate between the effects resulting from the
patient being subjected to a strange environment
and the effects which may result because of
exposure to pollutants. In addition to the
standard medical clinics, **special clinics are
frequently set up for pollution studies.** Examples
of this are the submarine room atmosphere used to
examine children (this room was a duplicate of the
Nautilus submarine interior) and automotive and
paint fume smog chambers to study eye irritation.

Experimental studies may involve either man
or other types of animals. When man is the subject,
the conditions must be controlled very percisely
and specific limitations are imposed by the United
States Department of Health (2). When other types
of animals are used for study, it is necessary to
attempt to relate effects caused by the pollution
in the test animal to the effects that may occur
in man. Problems exist when trying to relate data
from animals that have no diaphragms and may have
different heart beat rate, body temperature, type
of blood flow, and feather or dense hair covering.
In addition, if the animal is out of its natural
environment, this added stress must be compensated
for when the data is analyzed.

Bees are an example of sensitive animals used
for air pollution studies. The effects of bee
activity, honey production and water intake are

all indicative of the animal's health. Coturinex
quail (Japanese quail) are useful subjects for
air pollution studies because they mature in 35
days, they eat and drink copiously and they breed
easily. Guinea pigs and dogs are common
experimental animals because they are cheaply
obtained and readily available. When it is not
possible to choose an experimental test animal
with physical systems identical to man, animals
with systems as similar as possible must be chosen,
and the differences must be accounted for.

All of the effects during an experimental
study must be evaluated. For example, animal
waste products contain ammonia. If a test animal
is subject to an acid type of pollutant, such as
$SO_2$, and waste is allowed to accumulate, the
ammonia can neutralize the $SO_2$ and reduce the
actual concentration of **pollutant.**

It is possible to subject an experimental
animal to either acute or chronic concentrations
of a pollutant. Acute exposure is subjecting to
high concentration for short periods of time.
$LD_{50}$ (lethal dose where 50% of the subjects die)
are not normally used in air pollution studies.
Chronic exposure to low concentrations over long
periods of time so that only the normal death rate
occurs during the time of the experiment is the
normal test procedure.

It is frequently desirable to introduce
variations in the experiment. Sensitivity of the
test animal can be increased by adding stresses in
addition to the pollutants. For example: the
animals may be made to exercise; the temperature
may be increased; the humidity may be elevated or
depressed; etc. Another variation is to choose a
more sensitive test animal--young animals or
embryonic tissue are more susceptible to pollutants.
Sick animals and old animals are not recommended
for experimental studies because their death rate
is higher because of these factors and would confuse
the data obtained (3).

## 5.2.1 RESPIRATORY SYSTEM

Both gaseous and particulate pollutants are
drawn into the respiratory system. The respiratory
system attempts to prevent both types of pollutants
from reaching the lungs by various defense
mechanisms. The gas defense is the weakest and
consists of absorbing the gases with the moist

mucous that coats the respiratory tract.  What
happens to the gases that reach the lungs depends
upon the solubility of the gases as to whether
they can enter the blood system.
    The respiratory defense for elimination of
aerosol type particulate matter is more elaborate.
Figure 5.3 shows the human respiratory system.

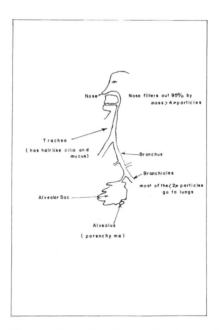

'Fig. 5.3.   Human Respiratory System--Aerosol
            Removal Defense

The nose hair serves as a first line of defense and
filters out the larger aerosols--95% of the particles
greater than 4 micron diameter are removed by the
nose hair.  However, essentially none of the less
than 1.2 micron particles are removed by the nose
hair.  The trachea has hair-like cilia which move
upward continuously in a peristalsis type of motion.
The rate of this upward motion is 3 to 4 centimeters
per second.  The bronchus also has cilia which move
upward at a rate of 0.15 cps.  Aerosols can be
deposited on the walls of the trachea and bronchus
by impaction, interception, sedimentation, or

Brownian diffusion. The first three mechanisms
tend to dominate and remove the larger than two
micron particles. Brownian motion influences the
small particles, however, there is not adequate
time for deposition of these particles during
their passage through the respiratory tubes. The
bronchioles connect the alveolar sacs with the
bronchus. The gas exchange between the lungs and
the bloodstream takes place in the alveolus
(parenchyma) which are around the edges of each
alveolar sac. Up to 75% by mass of the particles
entering can be deposited in the respiratory system
with up to 12.5% deposited in the lungs. The lung
deposits consist mostly of the less than 2 micron
diameter particles. In passing, it can be noted
that most of the aerosol called "smoke" consists of
less than two micron particles and therefore, a
great deal of this material ends up in the lungs.

## 5.2.2   RESPIRATORY DEPOSITION FACTORS

Deposition of aerosols in the lungs includes
physical, chemical and physiological factors.
Physical deposition factors are perhaps the most
important group and can be broken down into the
following five sections. 1) Gravitational
settling influences particles greater than 2 microns
in diameter and causes them to move downward in the
direction of the lungs (when man is in a normal
upright position). An aerosol particle in the
respiratory system must move against gravity or it
will become deposited. 2) Kinetic impact removal
occurs when the particle is directed toward a
target at a high velocity. The target in the
respiratory tract is either nose hairs, the cilia
or membrane walls. Frequently, particles that strike
a target do not adhere to the target and are re-
entrained by the gas flow. Adhesiveness of
particles to the respiratory tract is best for
particles less than 1 micron in diameter. 3) Viscous
force of the respiratory medium influences
deposition. The viscous resistance is proportional
to viscosity, velocity of the gas and aerosol
diameter. The thicker the mucus becomes, the
greater will be the deposition due to the viscous
force. 4) Brownian movement is the diffusional
movement of the particles in the gas which is
similar to molecular diffusion. This motion
causes particles to strike surfaces where
deposition can occur. Brownian movement is most
effective for particles less than 0.1 micron in

diameter.  5) The remaining factors of deposition
are:   agglomoration of the particles (which is
proportional to the concentration squared);
electrical charge on the aerosol which can cause it
to be attracted or repelled from a deposition
surface; condensation of a vapor which creates
aerosol particles that can be deposited; and
thermal force which is proportional to the temperature difference and particle diameter.  In addition,
chemical reactions can also influence the deposition
of particles.  Some of the particles react with
the mucus or the tissues making it easier for
either the particle or the reaction product to
become deposited or absorbed in the mucus.

Physiological factors such as tidal volume
and frequency of breathing influence deposition.
Both increased tidal volume and increased
frequence of breathing increases the number of
particles deposited.  Theoretically, there is an
optimum condition of tidal volume (See Appendix A)
and breathing frequency which gives the minimum
number of particles deposited for any particular
individual.  Data on this are not available,
although the National TB Association notes some
suggested breathing habits (4).

## 5.2.3   RETENTION OF POLLUTANTS

Pollutants that are deposited in the respiratory
system can be either eliminated or retained in the
body.  Elimination mechanisms depend upon where
the particle is deposited.  Particles in the nose
can be blown out or swallowed.  Aerosols that are
deposited in the upper respiratory tract can
either flow by gravity to the lungs or can move
upwards in the mucus by the peristalsis action of
the cilia.  Particles that reach the paranchema
region of the lungs have three alternatives:
1) they can become deposited in the lungs and stay
there permanently or chemically react with the
lung tissue--the reactions products can then
either stay there or be removed;  2) they can be
absorbed and enter the circulation system by
passing through the lymph cells into the blood; or
3) they can be coughed up into the upper
respiratory tract.

Emphysema is a chronic lung disease for which
there is no cure and is caused or at least aggra-
vated by air pollution.  Emphysema occurs when the
alveoli of the lung walls break down, reducing the

amount of membrane available to carry out the oxygen transfer. In addition, there is a narrowing of the smallest branches of the bronchioles which further restricts air exchange. Figure 5.4 is a schematic representing (A) the alveoli in a normal section of lung and (B) alveoli which have broken down and bronchioles are restricted due to emphysema.

A working man can breath in 125 liters air/min. This amount of air would fill a 10 by 10 by 8 ft. room in 24 hours. If the air contained 75 micrograms/$m^3$ of dirt, this man would take into his respiratory system 0.0135 grams per day of dirt. If 75% of this dirt were retained in the respiratory system, the body picks up about 0.15 ounces of dirt per year breathing air at this indicated concentration. In addition, the body has the added task of removing pollutants that are absorbed or ingested. It shoud be obvious why the ambient air should contain as little dirt as possible.

## 5.2.4 GENERAL EFFECTS OF POLLUTANTS

We have already mentioned that pollution can enter the body through ingestion by the mouth (resulting in gastrointestinal absorption, etc.), inhalation through the respiratory system where it can enter the body blood chemistry and by absorption through the skin where it can also enter into the body blood chemistry. Once in the blood, the pollutants may be absorbed by body tissues or may be deposited in the body organs. The most susceptible organs are the liver, kidneys, lungs, heart and brain.

Superficial irritation by pollutants of the skin, eyes and respiratory system may occur immediately upon contact with the pollutant. The damage also may be more extensive and not become evidenced for a relatively long period of time. Phenol, for example, irritates the skin. However, it is also absorbed into the blood stream and destroys body tissue. Absorption of sufficient amounts causes death in a few minutes. Beryllium oxide, on the other hand, may become deposited in the lungs (or anywhere else in the body) and remain there for years before damage is noticed. During this time, the beryllium oxide irritates the surfaces contacted by mechanical abrasion.

It is often possible for the body to clean itself by the normal elimination process and

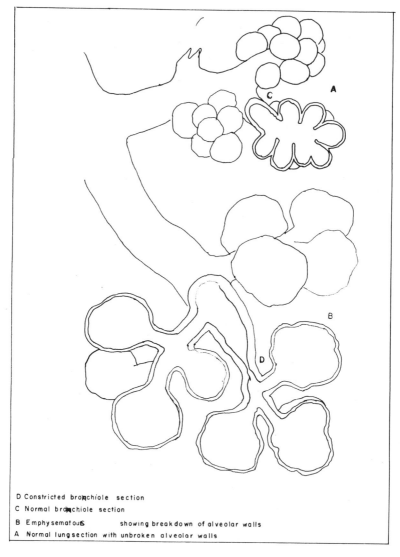

D Constricted bronchiole  section

C Normal bronchiole  section

B Emphysematous          showing breakdown of alveolar walls

A Normal lung section with unbroken  alveolar walls

Figure 5.4    Comparison of Normal and Emphysema
Diseased Lung Sections

thereby eliminate the pollutants or the reaction
products from the pollutants.  The damaged tissues
and organs then may (or may not) recover from the
pollution damage.  Organs such as the liver, which

purify other parts of the body, can become poisoned themselves during the clean-up operations.

It is also possible for pollutants to damage the body by only temporarily displacing some other substance. Binoxia, which is death from lack of oxygen, can be caused by the presence of excessive amounts of simple asphyxiants (such as carbon di-oxide or methane) in the air. When this happens, the lack of oxygen destroys brain cells and slows down the central nervous system. Table 5.1 shows how some atmospheric gases are distributed in the various parts of the body as calculated using Dalton's law and by actual measurements.

TABLE 5.1   GAS DISTRIBUTION IN THE BODY AT NORMAL PRESSURE

| | ATMOSPHERIC AIR | | PARTIAL PRESSURE, mm Hg in: | | |
| GAS | % | pp,mmHg | LUNGS | BLOOD | BRAIN CELLS |
|---|---|---|---|---|---|
| $O_2$ | 21.0 | 155 | 100 | 95 | 40 |
| $CO_2$ | 0.04 | 0.3 | 48 | 40 | 45 |
| $H_2O$ | 0.75 | 0.75 | 47 | 47 | 47 |
| $N_2$ | 78.9 | 573 | 517 | 572 | 577 |
| Argon | 0.93 | 7.1 | + | + | + |
| Others | "Small" | + | + | + | + |

5.2.5   SPECIFIC POLLUTANTS

Some of the more detrimental atmospheric pollutants and their effects on man are discussed in this section. Complete data relating ambient atmospheric concentrations and adverse effects on man are not always available. When such is the case, the industrial threshold limit values (TLV) set by the American conference of Governmental Industrial Hygienists will be indicated. The TLV was formerly known as the maximum allowable concentration (MAC). It must be remembered that the industrial threshold values are designed for use when the exposure time is a maximum of eight hours per day and six days per week. It is then necessary to relate these values to the maximum ambient air concentration which a person may be subjected to for up to 24 hours a day and seven days a week.  A rule of thumb used in this section when needed to convert industrial threshold values to ambient air concentration maximums is to use one tenth the TLV.  References 5, 6 and 7 are valuable guides for obtaining toxicity data and industrial TLV.

The Feneral Air Quality Standards (Code of Federal Regulation, Title 42, part 410 of Chapter IV, April 30, 1971) set forth some Primary Standards for ambient air quality to protect health which must be met by July 1, 1975, and some Secondary Standards for ambient air quality to protect welfare which must be met by October 1, 1977.

*PARTICULATES*

The Federal Primary Standard maximum for all particulates combined is 75 $\mu g/m^3$ annual geometric mean (260 $\mu g/m^3$ 24 hour max, once per year) and the Secondary Standard maximum is 60 $\mu g/m^3$ annual geometric mean (150 $\mu g/m^3$ 24 hour max, once per year).

Antimony--Antimony from metal, oxides or other compounds damages the heart as evidenced by data from electrocardiograph readings.  This damage appears to be irreversible.  Antimony also damages the kidneys by causing tubular cell damage which closes the tubes.  As a result, the body electrolyte concentration and water retention increases.  Blood pH goes from 7.2 to 7.6.  The industrial threshold is 0.5 $mg/m^3$ making the suggested annual geometric mean maximum value for ambient air 0.05 $mg/m^3$.

Arsenic--Arsenic is a silver crystalline metal that exists as either a black or yellow allotrophic form of arsenic oxide.  The yellow arsenic and arsenic compounds are highly toxic.  Arsenic causes digestive system irritations resulting in stomach and intestinal upsets.  The recommended maximum concentration is 0.012 micrograms per cubic meter (8).

Asbestos--Asbestos is a group of magnesium silicate minerals which exists in fibrous form.  Asbestos causes a lung disorder called asbestosis which is similar to that of beryllosis or silicosis.  Prolonged exposure to asbestos results in shortness of breath, coughing and chest cancer.  It eventually can effect the hear and can cause clubbed fingers (clubbing of fingerprints occurs in heart and lung diseases and is a rounding and broadening of the fingertips). The recommended maximum ambient concentration of 4 million particles per cubic meter is suggested (8). Asbestos is a cumulative poison and persons exposed to asbestos are more subject to respiratory disorders.

Beryllium--Most damage due to beryllium is caused by beryllium oxide (BeO) which is an extremely chemically inert compound. Beryllium oxide dust, which has the appearance of chalk dust, is a cumulative poison because of the fact that the body cannot chemically decompose it. Acute damage causes beryllium disease which is beryllosis of the lungs due to the deposits that result in the lungs and lower respiratory tract. Ciliary action is stopped at concentrations of 1 mg/m$^3$. This results in lowered body resistance to respiratory diseases such as pneumonia and emphysema. Chronic beryllium poisoning may not be evidenced immediately and a biopsy (removal of tissue) is usually necessary to be sure. A latent period of five to fifteen years may exist before the poisoning systems are evidenced. The symptoms are pulmonary hypertension (high blood pressure in the lungs) and damage to the liver and kidneys. This is believed to upset the nitrogen balance in the body causing increased nitrogen brain legions to be formed. It also causes a decrease in tidal volume. The recommended annual geometric mean maximum of 0.0013 µg/m$^3$ of beryllium is suggested (8) and the proposed nation standard is 0.01 µg/m$^3$ average for 30 days.
    Beryllium oxide allowed to become deposited under the skin or in the eyes or nose will remain there and form an ulcer. Cancer of the lips and mouth are also reported due to contact with beryllium oxide. Other forms of beryllium (e.g. beryllium acetate) can also be poisonous.

Cadmium--Cadmium is a soft white ductile metal which turns grayish upon oxidation in the air. The metal can become tarnished and cause food poisoning and the oxide powder in the air is toxic causing lung damage. This results in shortness of breath and pain in the chest. It is also suspected of affecting the central nervous system and causing profuse sweating. A threshold value of 0.1 mg/m$^3$ is suggested indicating that the maximum ambient concentration should be 0.01 mg/m$^3$.

Lead--This includes lead, the lead oxides (Pbo, $Pb_2O_3$ and $Pb_3O^4$), as well as the lead salts (PbClBr and $PbNH_2Cl$) which originate from automobile exhausts. Lead is a cumulative poison. It damages the nerves causing deadening of the nerve sense receptors. A burning feeling of the feet is noticed in cases of lead poisoning. It also causes anemia which is a deficiency of red blood cells, and therefore, prevents absorption of vitamin $B_{12}$ resulting in malnutrition. Lead poisoning also causes bleeding. Probably no other chemical has a greater compilation of toxilogical literature than does lead.

Lead compounds can be inhaled. this makes it extremely important that the tetraethyl lead and tetramethyl lead additives in gasoline should be removed to prevent air pollution of lead due to auto exhaust. CO and lead are both products of auto exhaust and atmospheric measurements of these materials show a linear relation. Concentrations as high as 18.4 $\mu g/m^3$ of lead are observed along certain California freeway interchanges. Atmospheric lead concentration is lowest in residential areas--away from the freeways.

Lead can be absorbed through the skin and by ingestion. Air pollution deposited on plants which are eaten by foraging animals can, in turn, be eaten by man resulting in lead poisoning to man. Man also can eat the lead poisoned plants. Other foods that can contain lead include water and beverages. Tobacco smoke is known to contain lead and chewing on articles containing lead based paints can result in lead poisoning.

An annual geometric mean maximum of 1.0 microgram per cubic meter of elemental lead is suggested (8).

Mercury--Volatilized mercury is poisonous when inhaled. Elemental mercury is also irritating to the skin and cause a rash and swelling. Inhaled mercury results in damage to the gums. Most of the mercury inhaled is excreted in twelve hours, however, it does accumulate sufficiently to cause tingling of the extremeties, hearing difficulties, tunnel vision and (in extreme cases) eventual blindness. The threshold value of mercury is 0.1 $mg/m^3$ so the suggested maximum ambient concentration is then 0.01 $mg/m^3$.

Silica or Silicon Dioxide ($SiO_2$) forms silicic acid in the body. This acid destroys lung tissue and produces a fibrous growth in the lungs. The disease

silicosis results in shortness of breath, decreased
tidal volume and added stress on the heart. Persons
with silicosis are more susceptible to respiratory
diseases such as pneumonia, tuberculosis and
emphysema. The nodular fibrosis created in the
paranchyma region of the lungs has been observed to
continue to grow for several years after termination
of exposure. The industrial threshold limiting
value is 0.7 billion particles/m³, making the suggested
maximum ambient value 70 million particles/m³.

Sulfuric Acid Mist (H₂SO₄) damages the bronchus and
bronchioles as well as the lungs. It may cause
spasms in the subject at high concentrations. The
effects of sulfuric acid mist are in general, similar
to those of sulfur dioxide gas. A maximum annual
geometric of 4 µg/m³ is suggested (8).

*GASEOUS POLLUTANTS*

Carbon Dioxide (CO₂) if considered as a pollutant,
would be the most abundant atmospheric pollutant.
Normal atmospheric concentration of carbon dioxide
is usually less than 200 ppm. Carbon dioxide in
itself is non-toxic and non-cumulative. The
poisonous potential occurs when it displaces oxygen
in the air, causing suffocation due to binoxia. At
5000 ppm (CO₂ in the air), brain cells are damaged
resulting in degradation of the central nervous system.
Table 5.1 is an example of the distribution of CO₂
and other gases in the body and in the air. This
threshold value suggests that a maximum annual geometric mean value of 500 ppm, which is 910 mg/m³ at
ambient conditions of 70°F and one atmosphere (8).

Carbon Monoxide (CO) when considered as a pollutant,
is the second most abundant atmospheric pollutant
in urban atmospheres. Carbon monoxide is not toxic
in itself and is non-cumulative. The dangers with
carbon monoxide occur because of the strong affinity
the hemoglobin has for carbon monoxide (affinity for
CO is 300 times that for O₂). This causes oxygen to
leave the tissues resulting in anoxicity. The effects
are headaches, loss of visual acuity and decreased
muscular coordination.
    The following indicates how much polluted air
must be breathed in (and therefore, how long it
takes) to saturate the blood with CO to a given
level. Normal blood contains about 20 cc of oxygen
per 100 milliliters of blood and there are about

5000 milliliters of blood in the body.  As the
carbon monoxide replaces the oxygen on the blood
hemoglobin, this is called saturating the blood.
Oxygen capacity of the hemoglobin is slightly over
20 ml O/100 ml blood.  The following calculation is
for 50% saturation:

$$(5,000 \text{ ml blood}) \quad \frac{20 \text{ ml gas}}{100 \text{ ml blood}} \quad (0.50) = 500 \text{ ml gas}$$

This shows that 500 ml of carbon monoxide gas taken
into the blood would result in 50% saturation of
the blood.  If air contained 700 ppm carbon monoxide,
then:

$$(500 \text{ ml}) \quad \frac{10^6 \text{ parts air}}{700 \text{ parts CO}} = 700 \text{ liters}$$

Therefore, 710 liters of air must be inhaled with
this CO concentration and if all the CO remains in
the blood, it will cause a 50% saturated condition.
If the alveouler ventilation rate is 3.5 liters per
minute, the time required to reach 50% saturation
is found by dividing the 710 liters by 3.5 liter per
minute, giving 200 minutes.  The actual degree of
saturation depends upon the original carbon monoxide
saturation in the blood as well as the concentration
of carbon monoxide in the air and the exposure
time.  A concentration of 100 ppm may or may not be
safe--depending on the individual, the length of
exposure and the initial burden.  People exposed
to high concentrations of CO by smoking (there is
up to 40,000 ppm CO in cigarette smoke) are already
burdened by an initial saturation of as much as 15%.
1000 ppm would normally cause immediate death after
a short, continous exposure.  Table 5.2 shows the
nervous system effects due to carbon monoxide as
expressed in percent saturation.
    The April 30, 1971 Federal Ambient Air Quality
Standard sets the 8 hour maximum for CO at 9 ppm
which is 10 mg/m$^3$.

Fluorides are both highly irritating and toxic.  A
buildup of fluorides in the body causes fluorosis
which is a fixation of the calcium in the body by
the fluorine.  The result is mottled teeth and
softened bones (bone density increase is frequently
reported).  There may also be calcification in some
ligaments.  Loss of weight, anemia and impairment
of growth is also observed.  The suggested maximum

TABLE 5.2   TYPICAL CENTRAL NERVOUS SYSTEM EFFECTS
            DUE TO CO
            *expressed in % saturation*

| Effects | Short Exposure | Long Exposure |
|---------|----------------|---------------|
| None | 0-10 | 0-10 |
| Slight headache | 10-20 | 10-20 |
| Collapse, coma, brain damage, death | 40-50 | 30-40 |

concentration of fluorides is 01001 ppm which is
0.8 µg/m$^3$ as hydrogen fluoride (8).

Gaseous Ions are groups of four to twelve molecules
of any of the normal atmospheric gases, such as
nitrogen, oxygen, carbon dioxide and/or water vapor
which cluster around an ion. These ion concentrations
are in the order of $10^6$ ions per cubic centimeter.
It is, therefore, possible to breathe in $5 \times 10^8$
ions per breath. Ions have a life of about four
minutes and a charge of approximately six electron
volts.
      The effects of aeroionization are not uniform.
However, it is known that man responds to both
negative and positive polarity ions. Ionization
of a localized atmosphere is not a cure, but an
effective method in treating hay fever, asthma,
burns and in postoperative recovery. The polluted
air in cities shows excessive amounts of large ions,
both positive and negative at the expense of the
smaller ones which are common in the clean, fresh
country air (9).
      Negative ions in the atmosphere appear to
improve mental attitude and promote beneficial
effects. Positive ions seem to depress mental
attitude. Even infants being treated for malnutrition cried less when exposed to light negative
ions than at other times. Further exploration is
necessary before this type of pollutant can be fully
discussed.

Hydrocarbons as described in Chapter IV are a vast
number of different compounds containing carbon
and hydrogen. Some hydrocarbons, such as methane
and acetylene, are simple asphyxiants and dilute
the air by removing oxygen to a level which is not
adequate to support life (similar to $CO_2$). Methane

typically exists in urban air at concentrations of
about 9 ppm.  Other hydrocarbons, such as anthracene,
are nontoxic but are carcinogenic and produce
cancers because of the impurities they contain.
Organic compounds which can be derived from hydrocarbons, such as phenol, can quickly poison the
body by affecting the central nervous system.  These
pollutants may be either inhaled or absorbed through
the skin.  Absorption of phenol has caused death in
as short a time as thirty minutes.  Prolonged
breathing of these types of compounds causes
digestive disturbances, difficulty in swallowing,
excessive salivation, nervous disorders and skin
eruptions.

With these few frief examples of hydrocarbons
and their affects, it is readily obvious why
concentration of hydrocarbons in the air should be
low.  The April 30, 1971 Federal Ambient Air
Quality Standard sets the maximum 3 hour non-methane
hydrocarbon concentration at 150 $\mu g/m^3$ which is
0.24 ppm as methane.  This limit is well below the
lower explosive limit of these gases so this type
of hazard will not be discussed.

Hydrogen Cyanide (HCN)--Sax (5) states that hydrocyanic acid and cyanides are true protoplasmic
poisons.  They combine in the tissues with the
cellular oxidation enzymes.  This keeps the oxygen
unavailable for use by the tissues and results in
death through asphyxia.  Hydrocyanic acid combines
with the methemoglobin which produces cherry-red
appearing venous blood.  Exposures as short as
thirty minutes can cause death.  If the subject
recovers, there is rarely any permanent disability.

It is suggested that the maximum atmospheric
concentration of HCN be 0.07 ppm which is 78.4
$\mu g/m^3$ (8).

Hydrogen Sulfide ($H_2S$) is as poisonous as hydrogen
cyanide.  It is both an irritant and an asphyxiant.
Hydrogen sulfide is an acid and reacts with the
tissues of the body.  It slows down the central
nervous system reducing reaction time and causing
headaches and diarrhea.  It has the ability to
either stimulate or depress the central nervous
system as well as the respiratory center.  Chronic
exposure can result in visual acuity and conjuncti-
vitis because of irritation to the eyes.

The suggested maximum concentration is 0.01
ppm which is 14.1 $\mu g/m^3$ of hydrogen sulfide (8).

The California Air Resources Board set their $H_2S$
standard at 30 ppb for one hour.

Nitrogen Oxides (NO) affect the body by irritating
the nose, eyes, and lungs.  Nitrogen dioxide can
be smelled at about 3 ppm.  From short time exposure:
nose and eye irritation begins at 10 ppm; chest
discombort is noticed when the concentration reaches
25 ppm; focal pneumonitis occurs when the concentration is from 50-100 ppm, and death occurs when the
concentration is greater than 500 ppm.  Nitrogen
dioxide is suspected of accelerating tumor growth
and decreasing the resistance of the body to
diseases.  The April 30, 1971 Federal Ambient Air
Quality Standard sets the annual arithmetic mean
for nitrogen oxides at 100 $\mu g/m^3$ which is 0.05
ppm as $NO_2$.

Oxidants - Ozone $(O_3)$--It has been mentioned that
atmospheric oxidants can be primarily ozone.  Ozone
in the atmosphere alters visual acuity which prevents
the eyes from focusing properly.  It also increases
the calcification of bones resulting in premature
aging and depletes body fat.  Ozone affects the
lungs by reducing respiratory oxygen uptake and
reducing tidal volume.  It also collapses the
alveoulei.  It has been noted that ozone in the
concentration of 0.1 to 1.0 ppm oxidizes body
enzymes creating chemically active radicals.  The
April 30, 1971 Federal Ambient Air Quality Standard
sets the 1 hour maximum for photochemical oxidants
at 150 $\mu g/m^3$ which is 0.08 ppm as ozone.

PAN (Peroxyacetyl Nitrate and Peroxyacyl Nitrates
is a pollutant that affects vegetation more so than
man.  However, PAN does cause eye irritation and
increased respiratory airway resistance.  Both eye
and respiratory irritation have been observed at
concentrations of 0.5 ppm.  PAN is also a photochemical oxidant.

Sulfur Dioxide $(SO_2)$ appears to damage man by
increasing the hear rate, lowering tidal volume,
decreasing mucus flow, increasing eye blinking and
is also thought to affect the central nervous system.
Like many other pollutants, sulfur dioxide affects
young and old people and those with respiratory
problems more so than healthy individuals.  Also,
it is more detrimental when combined with
particulates than in clean air containing only the

sulfur dioxide pollutant. Larson (10) presents the
following equation relating deaths and the combined
effects of $SO_2$ and particulates:

$$D = 0.65 \ SP$$

where D is the number of excess deaths during an
air pollution episode and S and P are concentrations
of $SO_2$ in ppm and suspended particulates in $\mu g/m^3$.
The equation was derived from regression analyses
of the episode data from London and New York City
(1966).
　　The industrial limit is 1 ppm. Sulfur dioxide
odor is noticed at 3 ppm. From short time exposure:
bronchial constriction is observed at 5 ppm; mucus
flow increases at about 11 ppm; the upper respiratory
system becomes irritated at 20 ppm $SO_2$ and the
ciliary movement decreases at 25 ppm.
　　The April 30, 1971 Federal Ambient Air Quality
Standard sets the maximum annual arithmetic mean
for $SO_2$ at 0.03 ppm which is 80 $\mu g/m^3$.

Tobacco Smoke is included as a special pollutant
because of the extreme importance and concern of
this subject. Tobacco smoke contains carbon monoxide
(up to 40,000 ppm), carbon dioxide, mineral dust
(lead has already been mentioned), fly ash, nitrogen
dioxide, nicotine and formaldehyde as well as other
organic tars. Tobacco smoke has an apparent
advantage in that it smoothes the muscles of the
body. However, this is also a disadvantage as it
decreases ciliary action. Tobacco smoke affects
the central nervous system causing bronchial
constriction and irritates the lungs which results
in coughing and ultimately cancer. The United
States Surgeon General's Advisory Committee on
Smoking and Health has issued one of the very
numerous reports on smoking and health (11).

*RADIOACTIVE MATTER*

　　Before the subject of effects of pollution on
man can be concluded, it is necessary to mention
radioactive matter. This consist of gaseous and
particulate matter and/or energy. Radiation
converts water to hydrogen peroxide in the body,
which oxidizes the cells and can cause varying
amounts of damage (up to and including death).
Radioactive particles in the lungs have been shown

to cause cancer. Radioactive particles could
enter the bloodstream and damage the body at any
point.

5.3  EFFECTS ON MATERIALS

In addition to affecting living organisms,
air pollution also has effects on material objects--
these can be broken down into direct and indirect
effects. The indirect effects are those that are
caused by visibility problems. Visibility is
reduced due to the presence of light scattering
aerosol particles and moisture in the air.
(Correlations have been established between light
scattering and relative humidity for many types of
aerosols--aerosols of a more hygroscopic nature
can be identified by this technique.) Accidents
and near accidents result from poor visibility.
Direct effects include the mechanisms of
deteoriation resulting from contact of pollutants
with objects. The rate of deterioration by these
direct mechanisms are increased by the presence of
natural moisture, sunlight, temperature extremes
and temperature variations. The five direct
mechanisms are:
1)  Abrasion--resulting from physical wear on
the material by an abrasive pollutant.
2)  Deposition and removal--frequent cleaning
to remove deposited pollutants wears away the
object in addition to costing money for the
cleaning processes.
3)  Chemical attack--metals can become tarnished
by reactions with hydrogen sulfide, etched by
acid mist, and oxidized by atmospheric oxidants;
stones and masonry can be converted to salts by
atmospheric acids; textiles and cellulose materials
can be embrittled by chemical reaction with atmospheric acids; ozone oxidizes rubber and causes it
to become brittle so that it cracks when stressed.
4)  Secondary chemical attack--when
pollutants are absorbed, they can be converted into
harmful substances by the absorbing medium (an
example of this occurs when sulfur dioxide is
absorbed by leather--the SO is converted to
sulfuric acid--the acid deteriorates the leather).
5)  Electrochemical corrosion--soluble
pollutants dissolved in water increase the
electrical conductivity of the water which increases
the rate of ferrous rusting (oxidation).
Table 5.3 summarizes a few of the direct damages
inflicted on materials by air pollutants.

TABLE 5.3  DIRECT MATERIAL DAMAGE BY AIR POLLUTION

| Mechanism | Principle Materials Attacked | Damage Resulting | Air Pollutants | Aiding Natural Factors |
|---|---|---|---|---|
| Abrasion | Stone, masonry, metals, painted surfaces, ceramics | Scratching, wearing away, esthetics | Fly ash, dust, metal oxides | Wind, sun, mechanical wear |
| Deposition | All | Esthetic value lowered | All particulates | Wind |
| Chemical attack | Painted surfaces, textiles, metals rubber dyes, papers | Peeling, weakening, cracking, aging, esthetics | $SO_2$, $H_2S$, $O_3$, acids | Sun, moisture, temperature |
| Secondary chemical attack | Leather, building materials | Weakening, cracking, esthetics | $SO_2$, organics | Mechanical and physical wear, wind, sun, temperature |
| Electro-chemical corrosion | Metals | Oxidation weakening esthetics | Acids, salts | Moisture, sun temperature |

A comprehensive study entitled "The Beaver Report" was made in England to study the costs of air pollution effects. This report indicates that the annual cost in a typical metropolitan area due to air pollution may be expected to have the following proportional relationship:

| *Cost Item* | *$/yr, Millions* | *Approximate Percentage* |
|---|---|---|
| Laundry | 125 | 10 |
| Painting & decorating | 150 | 12 |
| Cleaning & depreciation of buildings (not including houses) | 100 | 8 |
| Corrosion of metals | 125 | 10 |
| Damage to fabrics | 263 | 21 |
| Loss of efficiency (e.g., operation of transportation facilities, steel rails, etc...) | 500 | 40 |

These figures reflect a dollar value assessment to certain losses, however these figures are, at best, only estimates. It is interesting that almost half of the total costs reported are due to loss of efficiency of mechanical devices. The report states that the total of $1.26 billion probably does not include most of the damage due to air pollution because it is not detected. The above report does not include any costs of health and medical expenses due to air pollution. Also, no loss figures are included to show plant and animal losses.

Theoretically, it should be possible to find a minimum net cost resulting from air pollution effects *and* air pollution control as indicated by the schematic of Figure 5.5. Someday we may be able to correctly assess the *effect* costs as well as the *control* costs so that we neither overcontrol nor undercontrol.

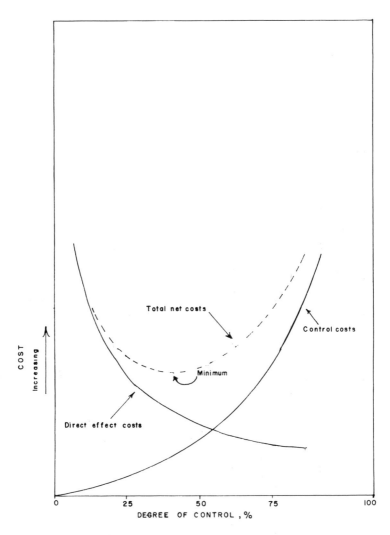

Figure 5.5   Generalized Air Pollution Costs vs
             Controls

## 5.4  EFFECTS OF PESTICIDES

Pesticides were developed specifically for the
purpose of improving the life of both plants and
animals (including man). Although this has been
accomplished to some degree, pesticides can also
be detrimental so it is appropriate to include
this brief discussion on pesticides.

Pesticides are chemicals designed to kill or
modify the growth of any pests (including insects,
weeds, fungi, rodents, etc.). In addition to
fighting pollutants, crops must compete against
50,000 fungi species, 30,000 weed species, 15,000
nematode (worms) species and 10,000 pest insects.
Pesticides are essential to farmers, public health
authorities, exterminators, gardeners and house-
wives, and when used sensibly and in moderation
serve important functions in the areas of food
supply and health. There are 20,000 people
employed in the production of these chemicals and
25,000 other people employed in the distribution
of them. The U.S. production of synthetic organic
pesticides is valued at over $1 x $10^9$/yr. in 1970
and this is predicted to increase about 10%/yr.
through 1975.

Pesticides have been effective in preventing
typhoid epidemics and in reducing malaria while at
the same time endangering health. In San Joaquin,
Bolivia, a DDT program was used to kill the malaria
carrying mosquitoes. The DDT also had the side
effect of killing the village cats. In the absence
of the cats, there was an invasion of a small
mouse-like mammal which carried the black typhus
virus. By the time it was discovered and the cats
replaced, there were 300 deaths due to the plague.
This is an example of an unusual situation which
resulted when chemicals were used without proper
precaution. Similar cycles, but usually of longer
time have occurred too many times.

Table 5.4 lists the six classes of pesticides
and shows their production in the United States in
1969 as well as the sales value received from each
class plus the average value received from the
sale of each compound of each class. Fungicides
are used to control scabs, rot and blight.

TABLE 5.4   U.S. PESTICIDE PRODUCTION IN 1969

| Classes | No. of Chemical Compounds | Product Sales Value $ MM/yr | Av. $ MM/yr per compound |
|---|---|---|---|
| Fungicides | 89 | 48.6 | 0.45 |
| Insecticides | 102 | | |
| Miticides | 26 | | |
| Nematocides | 13 | 237 | 2.0 |
| Rodenticides | 11 | | |
| Herbicides | 92 | 211 | 2.2 |

Insecticides are used mainly on crops and pests
while miticides are used in the control of spiders,
mites and slugs.  An example of the use for nemato-
cides is in the control of wire worms.  Most of
the rodenticides act as anti-coagulants in the
blood stream of rodents which causes them to
dehydrate and die.  Herbicides bring in the highest
value per compound and are used in the control of
weeds and grasses.
     The classes of pesticides may be further
subdivided according to the specific use of the
pesticide.  When used in closed buildings, such as
houses or grain warehouses, they are called
fumigants.  If they are used to control larva,
they are called larvacides.  Mollucides are used
to control snails and other mollusks.  Certain
chemicals have the ability to restore the pesticide
activity on resistance strains.  These are called
synergists, meaning that the effect is greater than
the sum of the individual effects.  Plant growth
regulators set blooms, thin fruit and prevent
lodging (falling over of long-stem grains).  If
the pesticide is used to repel insects, it is
called a repellant.
     More than 80% of the industry's output now
comes from organic chemicals.  Before World War II,
most of the pesticides were made from inorganic
materials.  For example, lead arsenate and calcium
arsenate were produced at the combined rate of
125 million pounds per year in 1940.  In 1965, only
11 million pounds were produced.
     Production of highly chlorinated organic
compounds was started at the end of World War II.
The production of these chemicals is on the
decrease.  In 1952, 100 million pounds of dichloro-
diphenyl-trichlorethane (DDT) were produced.  The
peak production of DDT was reached in 1962 with
167 million pounds.  The production for 1970 was
less than 65 million pounds and only a small
amount of DDT was produced in 1971 to fill the
vital but very limited use needs.  The benzene-
hexachloride (BHC) has now been essentially
replaced by the dieldren-heptachlor class of
compounds which appears to be non-toxic to man.
     The **highly** chlorinated DDT and BHC compounds
are not easily decomposed by nature.  This creates
the residue problem which now exists (approximately
2/3 of the DDT used has not decomposed by natural
chemical reactions).  DDT can accumulate in the
fatty tissues of animals.  Continued indiscriminate

use of DDT would have endangered people as well as
other organisms such as plankton, which are marine
animals that produce most of our atmospheric oxygen.
   New types of pesticides and new methods of
controlling the use of pesticides are continually
being developed. Organic farming is suggested by
many as a substitute for pesticides. This is the
use of organic matter (leaves, garbage, deactivated
human waste, etc.) and the selective use of plants
to control pests. This type of control is possible
for the small backyard gardener but it is not
practical at this time for commercial farming
operations. Another suggested alternative is
biological control of pests. This would affect
the hormone balance, especially of the juveniles
and cause the pest to die out. Use of insect
toxins and encapsulated bacteria is being studied
but these could be potentially dangerous to animals,
too. Natural predators are being used but must be
carefully watched to prevent any upset in the
ecological balance. Lady bugs, which prey on
insects and moths, and ichneumon flies, which
sting and eat other insects, are some of the
desirable predators.
   Use of chemicals in nature is a serious concern.
The National Chemical Institute reported in 1970
that 130 chemical compounds including 11 pesticides
can be sources of tumors in experimental animals
and they showed five insecticides, five fungicides
and one herbicide can cause tumors. Under the
Delaney Act, any chemical that can be shown to
cause cancer may be withdrawn from the market.
Continuous testing must be carried on so that the
chemicals used and their application methods are
both beneficial and safe.

SUGGESTED ADDITIONAL READING

   The vegetation damage handbook published by
The Pennsylvania State University (1) and the
pictorial atlas (12) prepared jointly by the Air
Pollution Control Association (APCA) and the
National Air Pollution Control Association (NAPCA)
are very complete information sources on this
subject. Both were written by specialists in this
field and intended to be used in the identification
of air pollution damage to vegetation. The ultimate
objective of the handbook is to obtain data for
determining annual loss in food and fiber production
due to air pollution damage. The atlas is the

result of a three year project and is intended as an information report.

The Sax text on dangerous properties (5) is actually a reference book giving toxicity data for an enormous number of chemicals. It also contains valuable sections on environmental quality, air pollution and radiation hazards.

QUESTIONS FOR DISCUSSION

1. Make a color sketch of a leaf showing the following air pollutant effects:  fluoride, $SO_2$, ozone, chloride and particulates.
2. Describe why the physical portion of the leaf attacked by the pollutants mentioned in No. 1 makes the damaged areas appear as you show them.
3. Discuss cumulative and non-cumulative pollutants as related to both plants and animals.
4. Estimated how much air pollution has cost you in the past year because of damage to (a) plants, (b) animals and (c) materials.
5. Describe how air pollution has affected you or someone in your family.  Include such related factors as health and concentration and type of air pollutants.  The effects can be either beneficial or detrimental and may be related to income or health.

PROBLEMS

5.1 Describe how air pollution quality and quantity could be monitored using only plants and/or animals (include limitations).
5.2 Conduct a field study in your area by obtaining interview data from ten individuals.  Include effects of air pollution upon these persons and their possessions.  Include where the damage occurred, when (including time of day) and include as many related environmental factors as possible. Correlate these results and attempt to establish what the air pollutants are, where they come from, their effects and what is and should be done about it.

REFERENCES

1.   Lacasse, J.L. and W.J. Moroz (editors), "Handbook of Effects Assessment - Vegetation Damage," Center for Air Environmental Studies, The Pennsylvania State University (1969).

2. Grad, F.P., "Public Health Law Manual," The American Public Health Association, Inc., Library of Congress, Card No. 65-26945 (1965).

3. "Introduction to Respiratory Diseases," National Tuberculosis and Respiratory Disease Association, Fourth Edition (1969).

4. Brecher, Ruth and Edward, "Breathing..What You Need to Know," National Tuberculosis and Respiratory Disease Assocation (1968).

5. Sax, N.I., "Dangerous Properties of Industrial Materials," Third Edition, Reinhold Book Corporation (1968).

6. Rose, Arthur and Elizabeth, "The Condensed Chemical Dictionary," Seventh Edition, Reinhold Publishing Corporation (1966).

7. "Hazardous Materials," Third Edition, National Fire Protection Association (1969).

8. "Rules and Regulations Governing the Control of Air Pollution, Chapter IV, Air Quality Standards," State of Illinois Air Pollution Control Board (submitted by TAC June 12, 1970).

9. Kornblueh, I.H., "Seminar on Human Bio-Meteorology - Air Ions and Human Health," U.S. Department of Health, Education and Welfare, Public Health Service; Publication No. 999-AP-25 (1967).

10. Larsen, R.I., "Relating Air Pollution Effects to Concentration and Control," JAPCA, Vol. 20, No. 4, pp. 214-225 (1970).

11. Stanhope, B.J. et al., "Smoking and Health," Public Health Service, Publication No. 1103, Second Printing (1965).

12. Jacobson, J.S. and A.C. Hill (editros), "Recognition of Air Pollution Injury to Vegetation: A Pictorial Atlas," Prepared under the auspices of APCA and NAPCA (1970).

# CHAPTER VI
# AUTOMOTIVE POLLUTION

It has been stated in Chapter II that exhaust
from transportation sources accounts for over 60%
by weight of the total pollution released into the
atmosphere in the United States. For this reason,
a separate chapter is devoted to these sources and
methods of reducing this pollution. Transportation
sources include all vehicles that use combustion
engines whether they move on the land, sea or in the
air. Farm, industrial and other work and pleasure
machinery are included in this discussion if they
are powered by internal combustion engines. Internal
combustion engines include turbine, jet, spark
ignition and compression ignition. As a general
rule, most of the smaller transportation vehicles
use the spark ignition or gasoline fueled type
engines. This includes most automobiles as well as
light trucks, tractors, small aircraft, boats,
lawnmowers,.... Compression ignition engines can
operate on gasoline, however, it is more economical
to operate them on the cheaper diesel fuel oil. The
oil fueled engines include compression ignition,
turbine and jet engines. So far, the turbine
engines have been used almost exclusively for medium
and larger size aircraft while compression ignition
engines are used on the larger trucks, tractors
and commercial land vehicles. Jet engines are
almost exclusively used on military, commercial and
executive aircraft.

The automotive engines used in private vehicles
produce most of the total atmosphere pollution
because of the great number of vehicles. This
chapter is devoted mainly to the automotive engine,
the related emission control activites and the
fuel it uses. It is not likely that the internal
combustion engine will be discarded in the near

future, yet it is necessary to reduce the amount of pollution from this device. Two approaches are discussed:   (1) modify the fuel and (2) modify the engine.

## 6.1   FUELS

### 6.1.1   CRUDE OIL

Crude oil is a petroleum product normally obtained from oil wells although a small amount is obtained from surface shale pits.  From underground and offshore wells, the United States produces about four million barrels of oil per day, making it the number two producer (the USSR is number one and Venezuela number three).  The United States produces 87.8% of its oil demands at a cost of $3/barrel--the rest is imported (limited by law to 12.2% in 1970) and costs 30% less.  The United States uses more than twenty barrels of oil per year per person (3364 x $10^6$ barrels in 1969 and estimated as 3980 x $10^6$ barrels/day in 1970) and is the largest consumer of crude oil with Canade being second.  Crude oil refining capacity is currently increasing about 6% per year.

There are three basic types of crude oil: paraffin base, napthene base and mixed base.  The paraffin base contains more open chain (or paraffin series type) hydrocarbons of both the saturated and unsaturated types.  The napthene base crude contains more cyclic or aromatic hydrocarbons while the mixed base contains a mixture of both open and closed chain hydrocarbons.  Variations in the type of crude result from the fact that the crude is obtained from geographically different wells.  For example, Pennsylvania wells produce more paraffin base crude while Texas and Louisiana wells produce more napthene base.  Illinois produces more of the mixed base crude.

Crude oil is processed at refineries to give the various petroleum products.  Currently, crudes are processed according to demands to yield about 56% gasoline (5.4 x $10^6$ barrel/day), 35% light oils, 7% heavy oils and 2% lubricants.  Light oils are mainly diesel and jet fuels, kerosene and heating oils.  Heavy oils are lubricating and asphalt oils. Lubricants include waxes and greases plus the heavy asphalts and tars.  Figure 6.1 is a schematic of a refining tower showing typical locations and temperatures at which various petroleum fractions are removed during continous fractionation of crude oils.

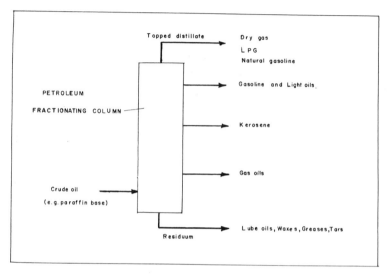

Figure 6.1   Crude Oil Fractionation

6.1.2   GASOLINE

Most gasoline is produced by cracking the gas oil fraction shown in Figure 6.1.  About 15% of gasoline exists naturally in crude oil, however, this is a poor quality material and must be blended with the artificially produced gasoline.  Motor grade gasolines are obtained by:   (1) straight distillation of crude oil, (2) cracking gas oil, (3) polymerization of $C_3$ and $C_4$ hydrocarbon gases, (4) natural gasoline from wet gas, and (5) from synthetic fuels (e.g. chemically processing coal).
Cracking of gas oils consists of thermally or catalytically breaking the longer hydrocarbon chains to produce shorter molecules ($C_6$ to $C_{11}$) of the type needed for gasoline fuels.  Some long chain hydrocarbons are also produced during the cracking process which results in the formation of tars.  When it is necessary to saturate the hydrocarbons and/or add hydrogen to the ends of the cracked chains, the refinery utilizes one of the numerous types of hydrogenation processes.  Polymerization consists of chemically combining short chains, such as $C_3$ and $C_4$, to form the hydrocarbons

with the necessary number of carbons required for gasoline fuels.

Gasoline is a mixture of a great many different kinds of hydrocarbons. Table 6.1 (1) lists most of these by group types and also shows typical variations in gasoline composition. Certain types of hydrocarbons make "better" fuel--for instance, olefins and branched chain paraffins give higher octane numbers.

*(Octane rating is an archaic method of gasoline grading which has been in common use since 1928. This consists of measuring the amount of audible knock produced in a standard test engine when iso-octane and normal-heptane are burned individually as fuels and comparing these results with the amount of knock obtained when a test gasoline is the fuel. Iso-octane fuel is given a knock rating of 100 octane and n-heptane is given a rating of 0. The American Society for Testing Materials (ASTM) test procedure D 908-55 specifies that research octane numbers be evaluated using an engine with a 3.25 inch bore and 4.50 inch stroke at various compression ratios from 4:1 to 10:1. The test is made at 600 rpm and ignition timing at 13⁰ BTC (Before Top Center) with intake air at 125⁰F. The current recommendations are to convert to the "motor octane" method which more closely approximates actual operating conditions. The ASTM test procedure D 357-53 for obtaining motor octane numbers uses the same engine at 900 rpm, an ignition timing dependent upon compression ratio and intake air from 75-125⁰F.)*

New types of engines such as the Wankel can operate on low octane gasoline (e.g. 66-67 octane), suggesting that fuels of the future may have a completely different performance criteria.

Refineries are able to raise the octane rating of gasoline without using lead by increasing the content of aromatic hydrocarbons and/or olefins. Restructuring of olefins in the gasoline to obtain optimum octane ratings can be performed by a process called alkylation which combines the light (short chain) olefins. It should be noted that although olefins can raise the octane number of gasoline, they are basic smog producing compounds as discussed in Chapter IV. Some states are attempting to restrict the amount of olefins in gasoline. Refineries also have the ability to produce more of the desired branch chain paraffins in the gasoline by isomerization which converts straight chains to branched chains. The use of isomerization (or other reforming processes such as cyclization, dehydrogenation, or cracking) requires very expensive catalysts containing small amounts of platinum and/or rhenium.

TABLE 6.1  COMPOSITION OF TYPICAL GASOLINES (1)

| Hydrocarbon | Mole percent in gasoline | | |
| --- | --- | --- | --- |
| | Gasoline A | Gasoline B | Gasoline C |
| Isobutane | 0.51 | 0.92 | 0.49 |
| n-Butane ¢ Neopentane | 6.83 | 17.49 | 6.27 |
| Isopentane | 11.07 | 0.73 | 0.74 |
| 2,3-Dimethylbutane | 1.77 | 0.73 | 0.74 |
| n-Pentane | 7.16 | 7.05 | 6.13 |
| 2-Methylpentane | 3.17 | 3.73 | 3.81 |
| 3-Methylpentane | 2.41 | 2.54 | 2.35 |
| n-Hexane | 4.24 | 4.56 | 3.95 |
| Methylcyclopentane & 2,2,3-dimethylpentane | 1.59 | 2.85 | 2.43 |
| 2,3-Dimethylpentane & 2,2,3-trimethylbutane | 1.22 | 0.47 | 0.26 |
| 2,3-Dimethylpentane & 2-methylhexane | 3.23 | 2.53 | 1.80 |
| 3-Methylhexane | 1.97 | 2.80 | 2.15 |
| n-Heptane | 2.15 | 1.65 | 1.79 |
| 3-Ethylpentane & iso-octane | 7.08 | 1.00 | 0.54 |
| Total Paraffins | 71.10 | 66.04 | 51.55 |

| Hydrocarbon | Mole percent in Gasoline | | |
|---|---|---|---|
| | Gasoline A | Gasoline B | Gasoline C |
| Benzene | 2.68 | 1.99 | 2.08 |
| Toluene | 5.94 | 7.52 | 6.30 |
| Ethylbenzene | 1.13 | 1.29 | 1.50 |
| m- and p-Xylene | 3.75 | 1.45 | 4.51 |
| o-Xylene | 1.40 | 1.73 | 2.21 |
| 3- and 4-Methylethyl-benzene | 1.18 | 1.39 | 1.41 |
| t- Butylbenzene & 1,2,4-trimethylbenzene | 1.31 | 1.51 | 1.40 |
| Total Aromatics | 20.00 | 20.44 | 24.25 |
| trans-2-Butene | 0.17 | 0.49 | 0.56 |
| cis-2-Butene | 0.10 | 0.36 | 0.55 |
| 1-Pentene | 0.30 | 0.32 | 1.12 |
| 2-Methyl-butene | 1.09 | 0.53 | 1.77 |
| 2-Methyl-1-pentene & 3-Hexene | 0.33 | 0.45 | 0.72 |
| 2-Methyl-2-pentene | 0.58 | 0.64 | 1.36 |
| 3-Methyl-cis-s-pentene | 0.35 | 0.71 | 0.66 |
| Cyclohexene | 0.14 | 0.32 | 0.55 |
| 1-Heptene | 0.09 | 0.20 | 0.66 |
| cis- and trans-3-Heptene | 0.59 | 0.95 | 1.16 |
| 1-Octane & 2-ethyl-1-hexene | 0.00 | 0.25 | 0.64 |
| 2,3-Dimethyl-2-hexene & trans-2-octane | 0.00 | 0.56 | 0.82 |
| Total Olefins | 8.90 | 13.52 | 24.21 |
| Total Hydrocarbons | 100.00 | 100.00 | 100.00 |

Numerous additives are put into gasoline in an attempt to improve the quality of the fuel. The most important, as far as air pollution is concerned, is the lead additives. Lead was introduced into gasoline starting in 1923 to make it burn more evenly and to prevent premature (spontaneous) combustion when the fuel is compressed in the engine cylinder. Each gallon of the 1970 gasoline contained about 2.5 grams of lead (6.3g tetraethyl lead). This amounts to one half million pounds of lead consumed per year in the combustion of approximately 83.5 million gallons of gasoline fuel. Up to 70% of the lead in gasoline is released with the car exhaust and it is estimated conservatively that half of this exhausted lead becomes airborne.

Figure 6.2 shows the increase in octane rating of regular and premium gasoline obtained by adding lead to refinery produced gasoline. In 1970, premium gasoline without tetraethyl lead at a normal refinery had an octane rating of about 94. Addition of tetraethyl lead increased the octane rating of this material to 99. Regular gasoline from the refinery has an octane number of about 86 and the tetraethyl lead increases this to approximately 94. Figure 6.3 shows that it would cost less than 1¢/gallon for a 90,000 barrel/day refinery to produce current octane quality gasoline without the addition of tetraethyl lead (TEL) (2). The dollar values on the chart are the total additional costs required at the indicated TEL level (2). Current refinery product costs are about 11¢/gallon for regular gasoline, 13¢/gallon for premium gasoline, and less than 10¢/gallon for either kerosene or No. 2 fuel oil.

The use of higher compression ratios in spark ignition engines makes it necessary to use gasoline with higher octane numbers. It has already been stated that octane numbers can be improved by various chemical processes in the petroleum refinery. This makes the addition of lead a matter of economic convenience. This would mean, however, that increased use of catalyst consuming operations such as catalytic reforming, hydrocracking, alkylation and isomerization would be necessary. This would put a severe strain on the demand for rare metal catalysts. Gasoline costs above refinery costs are due to delivery, handling and (mainly) to taxes. A recent survey in the New York City area showed that a 1¢/gallon variation due to processing catalyst and facility costs would scarcely be noticed. Lead-free gasoline has been made with

OCTANE RATINGS FOR

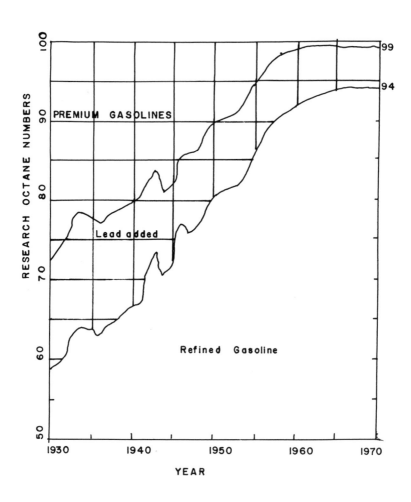

Figure 6.2a   Commercial Gasoline Octane Ratings

COMMERCIAL GASOLINE

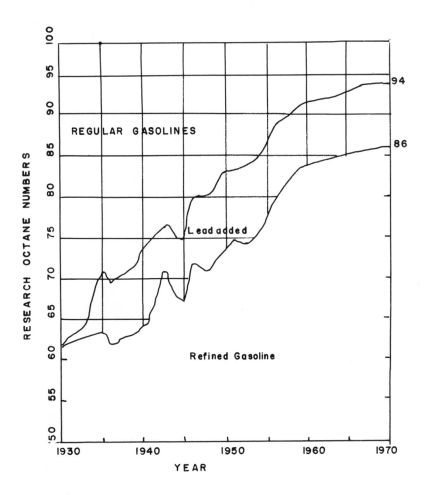

Figure 6.2b   Commercial Gasoline Octane Ratings

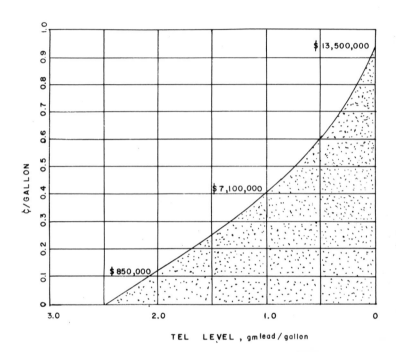

Figure 6.3   Estimated Cost of Lead Removal from
Gasoline ¢/Gallon vs Lead Level

high octane ratings by one petroleum company for
many years and sold as white gas (Amoco). The clear
lead free gasoline (when appropriately blended to
the proper octane number with aromatics) is up to
3% *more* dense than normal leaded gasolines. This
means that more BTU's are available per gallon
which should result in a net savings of about 1¢
per gallon.

Additives in gasoline can be summarized as
follows:

1. *Ethyl fluid*--This contains the lead added to
gasoline. The legal maximum is 4 milliliters per
gallon for automotive use. There is no restriction
for airplanes. It contains 63% tetraethyl lead
(TEL) or tetraethyl lead (TML), Ethyl fluid also

contains 26% ethylene dibromide and 9% dichloride
which act as lead scavengers.  Ethyl fluid also
contains 2% dye which includes some gum inhibiting
compounds.  This dye changes color if gums begin to
form in the gasoline.  The dye also serves to
indicate brand or type of gasoline.
2.  *Butane*--This is added to adjust vapor pressure
of the fuels.  Butane is a lighter, more volatile
hydrocarbon.  Winter fuels contain more butane for
ease in starting, and summer fuels should not contain
much butane, as it would vaporize and cause
excessive pollution.  The Reid vapor pressure test
is used to indicate the volatility (and, therefore,
the butane content) of fuels.  It consists of
adding chilled liquid gasoline fuel with four times
its volume of air to a pressure bomb, then measuring
the resulting pressure (corrected for air expansion)
at 100°F.  The state of California is attempting
to use a limit of 9 lb/in$^2$ to reduce excessive
volitization of gasoline in the summer.
3.  *Light lube oil* --These are added for upper
cylinder lubrication.
4.  *Tri methyl phosphate*--Lead compounds in the
fuel can build up across the spark plug and in the
combustion chamber.  The tri methyl (or cresyl)
phosphate is added to scavenge the lead and prevent
     plug fouling.  On the average, gasoline
contains 20 ppm phosphorous from this compound.
5.  *Other additives*--These include anti-oxidants
to prevent gum formation, rust inhibitors, anti-
icing compounds, etc.

## 6.1.3  OTHER FUELS

     Organic fuels are produced directly from fossil
fuel type of raw material.  In addition there are
other organic compounds which can be chemically
processed to yield automotive fuel.  Inorganic
fuels, such as ammonium nitrate, which are used
mainly as rocket motor fuels, radioactive fuel
sources and fuel cells are not considered in this
discussion.  In addition to kerosene and diesel
fuel oils, there are six commercial grades of
fuel oil.  The Grade 1 is the best grade of heating
oil with Grade 6 being the poorest quality fuel.
Diesel fuel oil is usually similar to Grade 2 which
is normally used for domestic heating.  The higher
the grade number, the thicker the oil because of the
increased content of higher chain, higher density
hydrocarbons.

The tendency for fuel to smoke is related to the hydrocarbon content as well as to the amount of oxygen present for combustion.  Assuming adequate oxygen is available, smoking increases in order with n-alkynes, iso-alkynes, olefins and aromatics.  Also, the shorter the length of the hydrocarbon chain, the more complete will be the combustion.  Liquified petroleum gas (LPG) consists of butane (mainly) and propane, and compressed natural gas (CNG) is primarily methane. These gases are stored under pressure as a liquid (LPG) or as a gas (CNG), then throttled (depressurized) to atmospheric pressure where they are used in their normal gaseous, rapid burning condition. Some of the low molecular weight alcohols are volatile liquids and can be used as low pollution fuels.

Explosives contain some or all of the required oxygen for combustion and because less nitrogen enters into the combustion reaction, they have the potential for creating less pollution.  Table 6.2 compares some of these fuels which could be used for spark ignition engines.  The lower heating value (LHV) means that some of the heat energy available in the fuel is used to convert water in the exhaust to a vapor.  The advantage of gasoline as a compact fuel source is evidenced by the LHV of the fuel at STP.  LPG, under pressure, will give almost as much available energy per unit of volume as liquid gasoline, plus it has the advantage of burning more completely.  The disadvantage is that a pressure vessel is required to contain the fuel. Fuels containing nitrogen have the potential disadvantage of producing more nitrogen oxides in the waste products.

The following petrochemicals produced from crude oil could be converted to gasoline, LPG, or any other fuel by proper chemical processing.  The values given in parentheses are in $10^6$ metric tons per year and are for the combined United States and Canadian production: ethylene (7.5), propylene, (3.6), butadiene (1.6), benzene (8.5), toulene (1.8) and xylenes (1.8) (source--ref. 3 and the Shell Chemical Company).

6.1.4  AIR-FUEL RATIO

An equivalent formula that represents the average of all the components in gasoline is

TABLE 6.2 POSSIBLE SPARK IGNITION ENGINE FUELS

| Fuel | Physical State in Combustion Chamber | Lower Heating Value (LHV) | | |
|---|---|---|---|---|
| | | BTU/ft | BTU/lb mole | BTU/lb* |
| Gasoline ($\approx C_8H_{17.5}$) | Liquid | / | 2,300,000 | 20,000 |
| Liquified Petroleum ($C_3H_8$) | Gas | 2,300 | 825,000* | 188,000 |
| Natural Gas ($CH_4$) | Gas | 1,000 | 344,600 | 21,500 |
| Hydrogen ($H_2$) | Gas | 285 | 102,136 | 51,068 |
| Blast Furnace Gas (60% $N_2$, 11.5% $CO_2$, 27.5% CO, 1% $H_2$) | Gas | 92 | / | / |
| Sewage Gas (68% $N_2$, 6% $N_2$, 22% $CO_2$, 2% $H_2$) | Gas | 621 | / | / |
| Alcohols – methanol ($CH_3OH$) – ethanol ($C_2H_5OH$) | Liquid Liquid | / / | 291,500 558,800 | 9,120 12,000 |
| Explosives – nitromethane ($CH_3NO_2$) | Liquid | / | 314,000 | 5,150 |
| – nitroethane ($C_2H_5NO_2$) | Liquid | / | 585,000 | 7,800 |

*At Standard Temperature and Pressure (STP) 32°F and 1 Atmosphere

$C_8H_{17.5}$.  Iso-octane, which is $C_8H_{18}$, is almost the chemical equivalent of average gasoline.  The stoichiometric equation for complete combustion of iso-octane with air, assuming only carbon dioxide and water vapor are formed by the combustion, is:

$$C_8H_{18} + 12\ 1/2\ O_2 + 47N_2 \rightarrow 8CO_2 + 9H_2O + 47N_2$$

Nitrogen is included in this equation because air consists of approximately 79% nitrogen and 21% oxygen by volume.  The air-fuel ratio (AF) is equal to the mass of air divided by the mass of fuel.  The AF ratio for stoichiometric combustion of gasoline then becomes approximately:

$$\text{mass of fuel} = (1 \text{ mole } C_8H_{18})\left(\frac{114 \text{ lb}}{\text{mole}}\right) = 114 \text{ lb}$$

$$\text{mass of air at SC} = (12 \text{ mole } O_2)\left(\frac{100 \text{ mole air}}{21 \text{ mole } O_2}\right)$$

$$\left(\frac{29 \text{ lb}}{\text{mole air}}\right) = 1730 \text{ lb}$$

$$AF = \frac{1730}{114} = 15 \text{ at stoichiometric conditions}$$

A rich mixture (low AF) means more fuel than stoichiometrically needed is used and results in more power for a given engine and also creates greater amounts of air pollution.  Lean mixtures (high AF) is when less fuel than stoichiometrically required is used; this produces a lower power output for the engine but gives less air pollution.

## 6.2  EXHAUST AND EMISSION STANDARDS

The exhaust from internal combustion engines can consist of products of combustion of the fuel with air (both complete and incomplete combustion products), unburned fuel and pieces of metal and metal oxides from the engine itself.  The following listing shows some particulate and gaseous exhaust components released by engines burning leaded gasoline:

PARTICULATES IN AUTOMOTIVE EXHAUST

| Element or Ion Identified | *Typical* (*) Weight Percent | *g/mi* | Remarks Related to These Components in the Exhaust |
|---|---|---|---|
| C | 28-34 | 0.115 | Mainly as soot and tars from incomplete fuel combustion |
| H | 6 | --- | As tars and/or water vapor |
| Fe | 1 | --- | Metal as oxides from engine rust, wear and chemical reactions. |
| Pb | 17-25 | 0.059 | As PbClBr halide or oxides from TEL or TML anti-knock compounds |
| Cl | 9 | --- | As lead halide |
| Br | 4 | --- | " " " (With both Cl and Br, PbClBr could equal 0.097 g/mi) |
| $H_2O$ | Varies | --- | Product of combustion; varies with temperature as to whether it is particulate or gas |
| S | --- | --- | Sulfur compounds including $SO_2$ and sulfuric acid mist are calculated as up to 0.1 g/mi (4) |
| $NO_3$ | 7 | --- | From gasoline additives |
| $NH_4$ | 5 | --- | " " " |
| Na & Ca | 3 | --- | " " " |
| Unknown | 12-14 | | Includes other metal oxides such as copper and aluminum from engine |

(*) Representing typical and weighted averages for cars using leaded fuel and tested with the Federal 7 mode cycle (4).

GASES IN AUTOMOTIVE EXHAUST   *(See Table 6.3 for amounts)*

| | |
|---|---|
| Hydrocarbons | Raw or partially burned gasoline; both long and short chain compounds; usually reported as methane |
| Carbon monoxide | Incomplete combustion product |
| Nitrogen oxides | Total nitrogen oxides; usually nitric oxide is a product of high temperature combustion |
| Carbon dioxide | Complete combustion product |
| Sulfur dioxide | Combustion product of sulfur in the fuel (low) |
| Oxygen | Excess or unburned air |
| Nitrogen | Excess or unburned air |
| Water vapor | Combustion product |

Data have been obtained showing that cars that use unleaded fuel emit about 40% less total particulates than cars that burn leaded fuel (4). However, the same data show that leaded fuel reduces carbon emissions during cold start and warm up conditions.

Automotive emission standards were finalized June 30, 1971. Table 6.3 lists the United States emission standards through the year 1976 and includes particulate standards and a 1980 goal. All values in the table except where otherwise noted are in grams of pollutant emitted in the exhaust per mile of travel of the vehicle. If the 1975-1976 standards shown are used, the emissions per automobile will be 90% less than those from 1970-1971 vehicles.

Factors which may influence the numerical values of the emission standards are the test methods. At the time the standards were established, emissions were measured using non-dispersive infrared analyses. The flame ionization detector is perhaps a more practicable test procedure for routine emission testing, but this new procedure reports higher values when compared with the former test method. In addition, the standards

listed in Table 6.3 were based on a chassis dynamometer simulated 7.5 mile urban trip beginning
from a cold start. The new test method proposed
for the 1973 vehicles weights emissions so that
43% of the cold start and 57% of the hot start
emissions make the composite exhaust sample which
more closely represents total actual air pollution
releases. If the new test is accepted, numerical
changes would be made to account for the testing
procedure and the 1973 values could be 28.0 g/mile
CO, 3.4 g/mile HC, and 3.1 g/mile $NO_x$.

Emissions from heavy duty gasoline fueled
engines are limited to 275 ppm HC and 1.5% CO for
1972 and later models. Heavy duty diesel engines
are limited to smoke emissions of Ringelmann #2 (40%
opacity) during acceleration and #1 (20% opacity)
during lugging.

Gaseous pollution measurements are made by
volumetric measurements and are included in Table
6.3 as ppm or volume %. The volumetric measurements
are converted to weight per mile measurements by
using a so-called "average" engine. The volume
of exhaust released by the average engine provides
the base for routinely converting volumetric
measurements (such as ppm) into total weight
emissions (as grams per mile). This is obviously
not an accurate method for establishing pollution
emissions. A small vehicle emitting a given volumetric concentration of pollutants will not
release the same total amount of pollutants per
mile as a large vehicle because of the smaller
exhaust volume. Therefore, the emission standard
indirectly specifies a pollution volumetric
concentration maximum.

It is difficult to measure the amount and
concentration of pollutants released by an automobile
in motion. The laboratory method is to totally
enclose a vehicle on a chassis dynamometer and make
a mass balance of the fuel and air input versus energy
and exhaust released. Emission tests run at manufacturers' plants or assembly factories are made on a
random basis. All new vehicles should undergo at
least an initial test and when suitable test
equipment is available used cars will have to be
subjected to routine inspection and emission
testing. The first state to set up official inspection stations to check air pollution control
devices was California. The Department of California
Highway Patrol published a handbook for inspection
stations in 1969 which describes methods of installing
and inspecting smog control devices (5).

TABLE 6.3   U.S. AUTOMOTIVE EMISSION STANDARDS AND GOALS

| Exhaust Pollutant | Car Model Year, Grams per Mile | | | | | |
|---|---|---|---|---|---|---|
| | 1970 | 1971 | 1973 | 1975 | 1976 | 1980 Goal |
| Particulates | / | / | / | 0.1* | 0.03* | 0.01* |
| Hydrocarbons (HC) | 4.1 (335 ppm) | 4.1 | 3.4 (275 ppm) | 0.41 (33.5 ppm) | 0.41 | 0.25* (8.4 ppm) |
| Carbon monoxide | 34.0 (1.45%) | 34.0 | 23.0 (1.0%) | 3.4 (1450 ppm) | 3.4 | 3.4 |
| Nitrogen oxides | / | 4.0* (1,100 ppm) | 3.0 (800 ppm) | 3.0 | 0.4 (110 ppm) | 0.4 |
| Fuel Tank HC | | 6g/day* | | | | |
| Carburetor HC | | 2g/soak* | | | | |

*Stated emission values--not standards (a soak is the 1 hour period after shut down at 180°F).

The hydrocarbon emissions in exhaust will vary with engine condition, type of engine, method of engine operation and type of fuel. Exhaust amounts and concentrations as a function of engine related variables are discussed in Section 6.4.2. The pollution emissions as related to the type of fuel have been discussed by Glasson (1) who shows that exhaust from different types of gasoline have a significantly different effect on photochemical oxidation. High olefin content gasoline will produce an exhaust with a similar composition--this olefin content causes faster photochemical reaction as indicated in Table 4.2.

## 6.3   SPARK IGNITION ENGINES

The fuel for spark ignition (SI) engines is usually gasoline. This liquid fuel is supplied to the engine by carburetion. Engines are essentially pumps, and as such, draw air into the combustion chamber during the intake stroke. The entering air is forced to flow through a restriction in the carburetor called a Venturi. At the narrowest point of the Venturi (tube), the pressure head loss is converted to an increase in kinetic energy head, causing a low pressure to exist at this position. Gasoline at the appropriate AF ratio is sucked through a calibrated nozzle into the tube because of this pressure differential and is either atomized or vaporized by the moving air stream. (Good atomization is essentially equivalent to vaporization at this point.) When the engine speeds up, more air and fuel are required. The increased pumping action of the engine causes air to flow more quickly past the Venturi tube which increases the pressure drop and brings in more fuel maintaining the desired constant air-fuel ratio.

Compression ratio is the clearance volume in the cylinder plus the displacement volume swept by the piston in one stroke divided by the clearance volume. (Incidentally, the clearance volume of the compressed gases is the volume of the combustion chamber.) The compression ratio was increased to a high of 9.5 which is the weighted average of the 1970 automobile engines. Increased compression ratio for an engine makes it possible to obtain more power from a given size engine. Figure 6.4 shows how power output expressed as indicated mean effective pressure (IMEP) has increased with

compression ratio over the years. As compression
ratio is increased, it is necessary to raise the
octane number to provide higher quality gasoline--
fuel that will not prematurely ignite due to auto
ignition. A compression ratio increase of from
five to ten results in an efficiency increase of
about 25%. At a compression ratio of ten, it is
theoretically possible to obtain 22 horsepower per
gallon of gasoline.

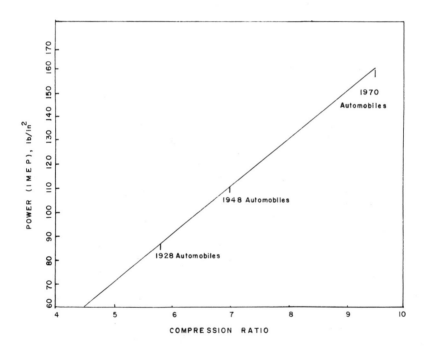

Figure 6.4   Spark Ignition Power Output versus
Compression Ratio

6.4   EMISSION CONTROL

6.4.1   FUEL CHANGES

It has already been discussed how various fuels
create significantly different air pollution
problems. One modification would be to make certain
that the exhaust contains no olefins. This could
be accomplised by removing olefins from the gasoline
used for fuel. In addition to varying the type of

gasoline, it is possible to change the fuel itself.
Liquid petroleum gas (LPG) and compressed natural
gas (CNG) fuel systems are increasing in
popularity. Spark ignition engines equipped with
these gas fuel systems do not have to meter and
atomize the fuel by carburetion as described for
the gasoline engine. The LPG system simply has a
control solenoid valve which throttles the fuel
(which turns to a gas after throttling) into the
combustion chamber at controlled times and time
intervals. CNG is stored in pressure tanks at up
to 2200 psig and it passes through a pressure
reducer before entering the control valve. Before
entering the combustion chamber, both systems usually
pass the fuel at about 0.07 psig through a special
air mixer filter. Cars equipped for dual LPG (or
CNG) and gasoline fuels have the mixer filter in
place of the standard air cleaner. No adjustments
to the engine are required to alternately use any
of these fuels. The CNG has an octane rating of
130 but a range of only 30-40 miles on a 140
pound tank.

These gas fuels keep the engine, spark plugs
and lubrication oil cleaner than gasoline. Data
obtained using LPG equipped automobiles (6) shows
that these vehicles produce less pollution than
present spark ignition engines: hydrocarbon
emissions ranged from 2.9 to 12.6 g/mile and $NO_x$
emissions ranged from 0.45 to 1.12 g/mile. Most
of these values are below the 1975 emission standard
requirements shown in Table 6.3.

Coal is composed mainly of napthenic aromatic
hydrocarbons which could be valuable as spark
ignition engine fuel. By destructive distillation,
disproportionation (molecular rearrangement) or
hydrogenation, coal can be converted to high
quality gasoline with high octane number and low
pollution potential. Drastic modifications in
type of fuel would necessitate engine changes.
This is discussed in Section 6.4.3.

6.4.2 ENGINE CHANGES AND ACCESSORIES

It is possible to reduce internal engine
emissions by modifying the engine so as to reduce
both the quantity and concentration of pollutants
emitted in the exhaust, crankcase, carburetor and
fuel tank. Even after engine emission improvements
are incorporated on new automobiles, and if they
can give some acceptable degree of continuous

performance, it will take a long time to lower the
total amount of automotive emissions because:
(1) automobiles in the United States are replaced
at a rate of only 10% per year and (2) the number
of automobiles in use increases each year.
    Figure 6.5 is a schematic showing where leakage
can occur in a typical automobile with no emission
controls.  Examination of this simple schematic
shows where controls should obviously be placed.
In the early 1960's, the draft tube which vented
the crankcase was recycled to the carburetor,
theoretically cutting down 20% of the hydrocarbon
emissions.  This is shown by a dotted line in
Figure 6.5.  From 1961 to 1963 and later, further
improvements were made when the engine itself was
vented to the carburetor by what is known as
positive crankcase ventilation (PCV).  This is also
shown by a dotted line in Figure 6.5.  The PCV
valve is a spring loaded air metering device that
opens when the pressure in the crankcase is greater
than the pressure in the carburetor permitting
hydrocarbon laden blow-by gases and vapors that have
leaked past the engine seals to be recycled through
the carburetor and into the combustion chamber.
The filtered air that is swept through the crankcase
comes from the air cleaner and is introduced through
the oil filter cap.  PCV valves become gummed and
stick, making it necessary to replace them
frequently (every 12,000 miles is recommended).
"Secondary air" is used on some automobile
engines from 1966 models on.  This consists of
forcing air into the exhaust manifold near the
exhaust valve.  The additional air serves to
promote complete combustion by providing excess air
to the exhaust gases, and it protects the valves
from overheating, but it causes more nitric oxide
in the exhaust (see Section 8.2).  This modification
requires an air pump to be effective.
    Evaporative losses have been reduced by
providing better quality gas tank caps.  In addition,
activated charcoal filters or traps are used to
adsorb evaporating material.  The double dotted
lines in Figure 6.6 show how evaporating gases can
be trapped by a charcoal canister.  Any gas that
passes from the canister to the atmosphere should
be clean and inert.  The solid lines show that
during engine operating periods, incoming air
reactivates the carbon in the canister by sweeping
the adsorbed gases back into the carburetor with
the air.  This system has the disadvantage of

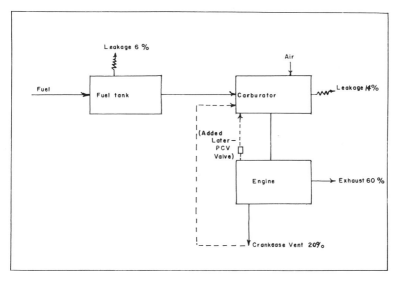

Figure 6.5    Automotive Hydrocarbon Emission Sources--
No Emission Controls

making too rich an AF ratio in the engine when it
is started, causing it to stall--especially if it
is warm.
      Other engine modifications have been made all
through the 1960's to reduce the amount of air
pollution.  Simple modifications were made to keep
the walls of the cylinder warmer because incomplete
combustion occurs at the cold cylinder walls.
Carburetion has also been modified to assure that
stoichiometric or nearly stoichiometric air fuel
ratio is used so that adequate air is available
for combustion and excessive hydrocarbons and carbon
monoxide are not released.
      Figure 6.7 shows how the exhaust composition
can vary as air fuel ratio (as measured at the
exhaust) is changed.  This figure shows two families
of curves.  The carbon monoxide and carbon dioxide
are read from the left hand ordinate which is in
percentage and the hydrocarbons and nitrogen oxides
are read from the right hand ordinate with values
expressed in ppm.  It can be seen that the carbon
monoxide content drops off drastically and
approaches a very low value as stoichiometric air
fuel ratio is approached.  The hydrocarbons follow
a similar trend although they appear to reach a
limiting asymptote.  The nitrogen oxides, in contrast,

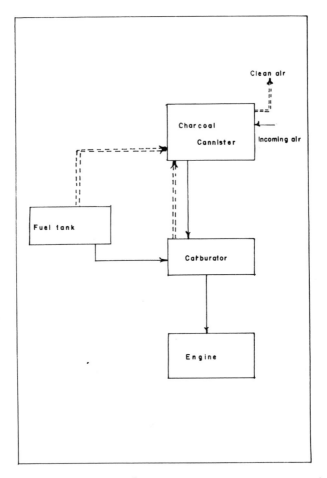

Figure 6.6   Carbon Adsorber to Stop Evaporative
Losses

increase in concentration and reach a peak at
slightly over stoichiometric air-fuel ratio.  Excess
air, then, is not advisable because of the increase
in $NO_x$ emissions.  Actual exhaust compositions will
vary with type of fuel, engine design, engine
condition, timing, type of pollution control
devices in use, the throttle setting and load on
the engine.

Exhaust studies made at the Southern Illinois
University in Carbondale, Illinois, showed that
essentially all 1970 model automobiles operate at

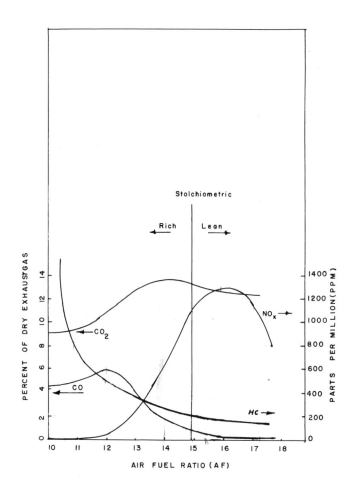

Figure 6.7  Typical Exhaust Gas Air Pollutants from
1971 Automobile (with Pollution Controls
Functioning)

an idle air-fuel ratio of approximately 14.2  Older
year model automobiles showed a variation in air-
fuel ratio, with one manufacturer (American Motors)
maintaining a significantly higher air-fuel ratio
than all other manufacturers.  This same study
showed that the cruise air-fuel ratio has been
increasing since 1964 (on all automobiles) from
13.5 in the early 1960 model cars to 14.2 in
1970.  The results of this study were in direct
agreement with the National Air Pollution Control

Administration findings (7). During deceleration,
the air-fuel ratio can drop to a low value (10-12)
resulting in extremely high hydrocarbon emissions
(from burning of the lubricating oil). During
acceleration, the air-fuel ratio usually increases
to about 15. As the cruise air-fuel ratio is
adjusted from 13 to 15, the fuel consumption
requirement decreases, but the power output also
decreases. The fuel consumption decrease is a
result of improved efficiency and the power loss
is a result of less fuel being consumed and not
available to produce energy.

Spark timing must be corrected for different
air-fuel ratios. A lean mix is slower burning than
a rich mix so timing must be retarded as air-fuel
ratios are increased. Figure 6.8 shows what might
happen at a given air-fuel ratio as the spark is
advanced. In general, a retarded spark gives less
hydrocarbons in the exhaust but it also reduces
engine output power.

Other accessories which are added to the exhaust
system are converters--including both catalytic and
thermal converters. These converters must serve
two functions: (1) to oxidize the unburned hydrocarbons to carbon dioxide and water and oxidize the
carbon monoxide to carbon dioxide, and (2) to reduce
the nitrogen oxides to nitrogen and oxygen gas.
Catalytic converters are capable of performing these
requirements; however, they are easily poisoned by
metals. Lead, which is contained in gasoline, is
such a poison. Results show that catalytic
converters work well when using unleaded gasoline.
Table 6.4 shows the results when a vanadium
pentoxide-copper oxide-aluminum oxide catalyst is
used with both leaded and unleaded gasoline. The
1971 model year cars had a lower compression ratio
(e.g., 8 to 8.5 as opposed to the 9.5 compression
ratio for 1970 cars). The reason for this **was**
to encourage the use of non-leaded lower octane
gasolines which can be currently produced without
extensive refinery modifications. This will help
by reducing the poisonous effects of lead on both
men and catalysts. Figure 6.9 shows how a typical
catalytic converter is poisoned by lead gasoline
preventing complete oxidation of both carbon
monoxide and hydrocarbons (2). (The lead content
of the gasoline used in this study was unspecified.)

Thermal converters are being produced that will
thermally oxidize hydrocarbons and carbon monoxide.
These thermal converters will also lower the amount

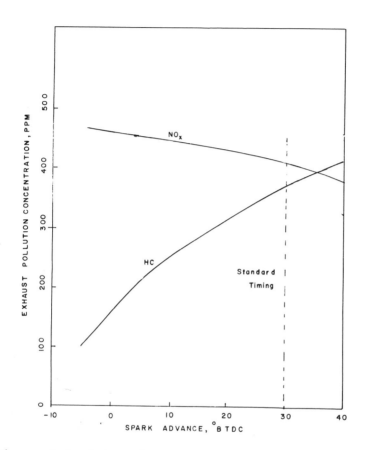

Figure 6.8    Auto Exhaust Pollutants as a Function
of Engine Timing (for a Spark Ignition
Engine at 30 mph Cruise)

TABLE 6.4    CATALYST POISONING BY LEADED GASOLINE

| | % of HC Removed After Miles | | |
|---|---|---|---|
| *Gasoline* | *0 mi* | *12,000 mi* | *18,000 mi* |
| unleaded | 80% | / | 70% |
| 1.5 ml TEL/gal | 80% | / | 29% |
| 3.0 ml TEL/gal | 80% | 0% | 0% |

of nitrogen oxides emitted by re-establishing a
chemical reaction condition favoring equilibrium of
nitrogen gas instead of the oxide.  DuPont claims
that an exhaust manifold thermal reactor tested in

100,000 miles of operation on a programmed chassis dynamometer has been able to reduce the emission levels of both hydrocarbons and carbon monoxides to below the 1975 federal standard (8).  These thermal reactors operate at $1650^{\circ}F$ under normal conditions. They result in a 5% loss in fuel economy and cost $150-$300.

The cost to the automobile owner for air pollution control devices for a ten-year period is listed below:

ESTIMATED COSTS TO AUTOMOBILE OWNERS FOR AIR
    POLLUTION CONTROL DEVICES (9)

| Year | Estimated Cost | Year | Estimated Cost |
|------|----------------|------|----------------|
| 1966 | $ 3.00 | 1971 | $ 40.00 |
| 1967 | 3.00 | 1972 | 40.00 |
| 1968 | 18.00 | 1973 | 100.00 |
| 1969 | 18.00 | 1974 | 120.60 |
| 1970 | 26.00 | 1975 | 314.00 |

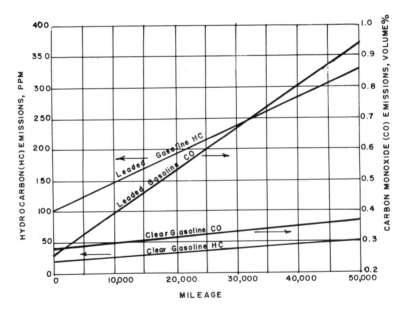

Figure 6.9   Typical Vehicle Emissions with a
             Catalytic Converter

An approximate breakdown by weight of the pollutant content of automobile exhaust is as follows: 0.4% particulates, 79.4% carbon monoxide, 5.4% nitrogen oxide, 14.4% hydrocarbons, 0.3% sulfur dioxide. Jet aircraft exhaust in comparison has the following approximate composition: 10.4% particulates, 22.7% carbon monoxide, 6.6% nitrogen oxide, 57.6% hydrocarbons and 2.8% sulfur dioxide. This shows that jet exhaust is about 42 times more "dirty" in particulate matter than automobile exhaust and also has much more hydrocarbons, but the amount of carbon monoxide is substantially less. This is due to the fact that the jet aircraft is a continuous burning, 'ho pumping' type of engine. No significant advances have been made recently in jet engine pollution control. The only appreciable work being done is to install after-burners on the engines which will burn the particulate matter (mainly carbon soot). This, of course, will produce more of the undesirable unburned hydrocarbons, carbon monoxide and nitrogen oxides.

The compression ignition or diesel engines operate at higher compression ratios than the spark ignition engine. Typical compression ratios for the diesel engines are 15 to. 1. These engines have a positive displacement type of fuel injection and do not have a carburetion system. The engine pumps a constant amount of air per each intake stroke. The amount of fuel injected is varied by a rack setting on a pressurized injection spray nozzle. If the injection setting is adjusted properly, excess fuel cannot enter the combustion chamber and smoking does not happen. If excessive amounts of fuel are permitted to enter the cylinder, the diesel engine smokes. A smoking engine is able to provide increased power at a loss of economy.

Diesel engines are not inherently more powerful than gasoline engines. Their advantage lies in the fact that the diesel cycle, with its higher compression ratio, is more efficient (thermally), so they are more economical to operate. In addition, the positive timed fuel injection makes it possible to use the cheaper diesel oil. No significant advances have been made in diesel engine design other than the assuring that proper fuel injection is maintained.

6.4.3   CHANGING ENGINES

Numerous studies have been undertaken to find a low pollution engine. They have revealed the fact that so far the internal combustion engines are the most practical, efficient, and economical power sources for automobiles. William Lear, who has done much experimental work on modern steam engines, had decided that it was impractical to compete economically against the internal combustion engine and was only continuing his research on commercial size steam engines. He has recently reported that the screw compressor can be used as a non-reciprocating type of external combustion engine. Tests so far show that this steam engine can perform at over 10% total efficiency (which is the actual efficiency limit of most spark ignition engines), and should be capable of producing a maximum efficiency of 12%.

The electric engine has the disadvantage of producing ozone gas when the engine is switched due to the sparking created in the motor and contacts. In addition, these vehicles require recharging by connecting them to electric utility systems which may not have electricity available for this purpose. As long as fossil fuels produce most of the electricity, use of electric automobiles would shift much of the pollution control burden to public utilities. Little work has been done on such exotic types of fuel systems for automotive transportation as fuel cells and solar energy; however, they are potential future fuels.

### 6.4.4   OTHER METHODS

The most promising method for eliminating transportation pollution appears to be in the development in high speed mass transit systems ; yet, these systems are far behind in development. These mass systems will have to be economical, available and pleasing to the public. A few systems have been developed and are now in an experimental stage. They include a jet propelled research vehicle that rides on a cushion of air. This vehicle is guided along tracks supported by air cushions pressurized by air compressors. The speed range is estimated to be from 150 to 300 mph. The guideway is a flat concrete horizontal surface using inverted "U" supports. The driving power comes from electric motors (with the reaction rail in the guideway) propelled by a gas turbine. At this point, it appears to have the disadvantages of having noisy and difficult switching systems. A "super flywheel" power system is also in development.

The flywheel is of a radically new design which
makes it capable of being rotated at 50,000 to
90,000 rpm in a vacuum.  This system would rev-up
the flywheel before starting and use this stored
energy for propulsion.
      Also in development are several tube type
vehicle systems.  The devices travel in enclosed
subterranean tubes on wheels, rails or air cushions.
They appear to be expensive to construct but are
potentially more economical to operate.  They can
be guided and supported electro-magnetically in the
tube by superconductive magnets attached to the
vehicle.  These devices have a potential top speed
of up to 500 mph.  Most of the propulsion power
comes from linear electric motors using pneumatic
and gravity assisted acceleration and braking.
Evacuated systems have the added advantage of having
reduced air drag.  A disadvantage of these systems
appears to be in the fact that the intense
magnetic fields may cause damaging effects upon
the passengers or their belongings.  Also, the
tunnels are costly to construct.

QUESTIONS FOR DISCUSSION

<u>1</u>.  Name and discuss the various types of automotive
engines (e.g. gas turbine and Wankel) which are
being proposed as possible pollution free engines.
<u>2</u>.  Suggest and discuss the practicability of some
potential control devices that could be used on
automotive systems.  (For example, Venturi scrubbers).
<u>3</u>.  What will be the public reaction to some of the
currently proposed future transportation systems?
<u>4</u>.  How can automotive manufacturers and vehicle
owners be made to keep the pollution emission
controls in effective operation?

PROBLEMS

<u>6.1</u>  a. Consider gasoline to be $C_8H_{17.5}$.  Write the
balanced stoichiometric equation for complete combustion of this fuel with air.
      b. What is the air fuel (AF) ratio?
      c.  How many cubic feet of air are consumed per
minute for the burning of this fuel, if the fuel is
burned in a four-stroke spark ignition engine with
a volume of 343 cubic inches operating at 4000 rpm?
      d. How many pounds per minute of the various
air pollutants will be released into the atmosphere
by the engine operating at these conditions? (Assume

the engine to have a 9.5-1 compression ratio and
standard control devices in good operating condition
and to be using leaded fuel; assume complete
displacement of exhaust and neglect volume of fuel.
NOTE: Engine volume is equal to clearance volume
plus displacement volume; compression ratio is total
volume divided by clearance.)
6.2 Devise a test procedure to measure the air
pollutants released from the automobile described
in 6.1 under road conditions.
6.3 Would you recommend the use of a catalytic or
thermal converter to remove the nitrogen oxides
and pollutants created by incomplete combustion?
Qualify your answer.
6.4 How can coal be converted to gasoline? Are
there any advantages or disadvantages of this as far
as the consumption of our coal reserves and in the
potential pollutants created in the combustion of
this type fuel?
6.5 Estimate what happens to the exhaust of the
same engine in 6.1 when the car is decelerating.
How much of the various pollutants would be emitted
now in pounds per minute?
6.6 Repeat 6.5 but during cruise.

REFERENCES

1.  Glasson, W.A., and C.S. Tuesday, "Hydrocarbon
    Reactivity and the Kinetics of the Atmospheric
    Photo-oxidation of Nitric Oxide," Journal of
    the Air Pollution Control Association, Vol. 20,
    No. 4, pp. 239-243 (1970)
2.  Logan, J.O., and C. G. Gerhold, "Summary of
    Statements to House Subcommittee on Public
    Health and Welfare," Washington, D.C., 12 pages,
    March 20 (1970)
3.  Collingswood, P., "World Developments in Olefin
    Supplies," presented at ACS/CIC meeting in
    Toronto, Canada, May (1970)
4.  Ter Haar, G.L., et.al., "Composition, Size and
    Control of Automotive Exhaust Particulates,"
    presented as paper #71-111 at the APCA meeting,
    Atlantic City, N.J., June (1971)
5.  "Handbook for Installation and Inspection
    Stations," Department of California Highway
    Patrol (Official Motor Vehicle Pollution Control
    Device Stations) 1969
6.  "Emission Tests on Three LP-Gas Vehicles,"
    Project M205, California Air Resources Board
    (1970)

7. "Auto Emission Systems Fail to do the Job,"
   Environmental Science and Technology, Vol. 4,
   No. 4, pp. 279-280 (1970)
8. Mikita, J.J., and E. N. Cantwell, "Exhaust
   Manifold Thermal Reactors--A Solution to the
   Automotive Emissions Problem," presented at the
   68th Annual National Petroleum Refiners
   Association, San Antonio, Texas (1970)
   (Revised June, 1970)
9. Taken from the Semi-Annual Report by the
   Committee on Motor Vehicle Emissions of the
   National Academy of Sciences to the U.S.
   Environmental Protection Agency, January 1 (1972)

# PART II
# ENGINEERING CONTROL

# CHAPTER VII
# CLASSIFICATION OF POLLUTANTS

In Part I of this book, we defined air pollutants as including both particulate and gaseous matter. Particulate matter was further classified as that material which is either solid or liquid and from 1000 microns down to sub-micron in size. A special group of particulates is the aerosols which are the less than 50 micron diameter material. Gaseous pollutants are those materials which exist in the atmosphere as either gas or vapor. At standard conditions (SC--considered to be 70°F and one atmosphere unless otherwise indicated), air may be treated as an ideal gas. Gaseous pollutants in the air are usually extremely low in concentration, making it possible to also treat them as ideal gases even though vapors and some of these same gases may not be treated as ideal gases in high concentrations.

In common engineering practice, it is an "unwritten law" that *unless otherwise stated:* concentrations of *gases* are expressed by *volume* (volume ratio is the same as mole or pressure ratio for ideal gases); and concentrations of *solids or liquids* are expressed by *weight*. Characteristically, both particulate and gaseous types of pollutants exist simultaneously in the atmosphere, and frequently this causes a synergistic effect (where the resultant effects are worse than the sum of the individual effects). This can be true for mixtures of both types of pollutants as well as for mixtures of the same type of pollutants.

Chapter 7 deals with the classification of pollutants by describing and defining some of their physical characteristics. This is necessary before an adequate discussion of control theory and control equipment can be considered. Classification of pollutants by source is not included here as this is covered in Chapter 2.

## 7.1  PARTICULATES

Particulate matter is usually considered to be spherical in shape.  However, this is not always the case and there are times when necessary corrections will have to be made to account for variations from spherical geometry.  Several methods for conveniently describing and accounting for atmospheric particulate matter are presented in the following five sub-sections.

### 7.1.1  DESCRIBING BY SIZE

Particulate matter in nature often has a natural or log-normal geometrical size distribution. The most common methods of classifying particulate matter is by mass (which is proportional to volume), surface area or by number.  Liquid particulate matter is essentially spherical in configuration and presents no problem in this method of classification, and it is possible to easily convert from one type of classification to another.  Solid particulate matter may or may not be spherical in shape, so equivalent sizes are often assigned.  One such equivalent size is Stokes' diameter which is obtained by noting the diameter of a sphere (of the same density) in free fall, which falls at the same rate as a non-spherical particle.  This is observed in the laminar flow region where Reynolds' number is less than 0.2

$$(\text{Reynolds' number} = N_{Re} = \frac{Dv\rho}{\mu} ).$$

Various diameters and the mathematical procedures for obtaining the arithmetic mean diameters for these types are listed in Table 7.1.

Methods to describe and account for particulate matter in a gaseous medium can utilize one or more physical properties of the particles.  The following is an alphabetical listing of some particle sizing methods and the related physical properties used for the measurements:

Acoustics--particles $5\mu$ and larger can be accelerated by sonic energy and then either decelerated or impinged on a target to determine size as a function of particle mass.

Beta Radiation--a beam of electrons is attenuated by absorption and reflection by the particles and the medium.  This is effective for obtaining mass for $0.1\mu$ and larger particles.

TABLE 7.1 PARTICLE DIAMETERS AND MEANS

| Symbol | Name | Obtained by | Mean Diameter | |
|---|---|---|---|---|
| | | | Symbol | Obtained from n Particles by: |
| $d$ | number diameter | measuring | $\bar{d}$ | $\dfrac{\Sigma nd}{\Sigma n}$ |
| $d_v$ | volume or mass (wt.) diameter | the cube root of [sphere volume divided by $\left(\dfrac{\pi}{6}\right)$] | $\bar{d}_v$ | $\left(\dfrac{\Sigma nd_v^{\,3}}{\Sigma n}\right)^{1/3}$ |
| $d_a$ | surface or area diameter | the square root of [sphere area divided by $\pi$] | $\bar{d}_a$ | $\left(\dfrac{\Sigma nd_a^{\,2}}{\Sigma n}\right)^{1/2}$ |
| $d_s$ | Stokes' diameter | diameter of sphere of same $\rho$ and falling speed or $\left(\dfrac{d_v^{\,3}}{d_d}\right)^{1/2}$ | $\bar{d}_s$ | $\dfrac{\Sigma nd_s}{\Sigma n}$ |
| $d_x$ | Sauter (volume-surface) diameter | $\dfrac{volume}{area}$ | $\bar{d}_x$ | $\dfrac{\Sigma nd_x^{\,3}}{\Sigma nd_x^{\,2}}$ |
| $d_p$ | projected diameter | diameter of sphere having same projected area | $\bar{d}_p$ | $\dfrac{\Sigma nd_p}{\Sigma n}$ |
| $d_d$ | drag diameter | diameter of sphere of same $\rho$ and drag resistance | $\bar{d}_d$ | $\dfrac{\Sigma nd_d}{\Sigma n}$ |

Condensation Nuclei Counter--0.005 to 2μ particles
cause condensation of supersaturated vapors.  The
resultant condensation particles indicate the size
of the original particles.

Elutriation--particle mass and medium drag can be
effective in separating 5 to 100μ particles.

Electrostatic Forces--depending on the charge a
particle accepts, size separation is effective for
2 to 60μ particles.

Holography--laser light can detect and reconstruct
magnified images of particles larger than 0.1μ.

Impaction--particle mass is utilized to size particles
from 0.1 to 100μ diameter.

Light Transmission--the wavelength at which a
particle becomes extinct and ceases to transmit
light energy is related to particle size.  This is
effective for particles 0.1 to 2μ in diameter.

Microscopy--particles down to 0.5 and 0.05μ
respectively can be sized depending on the particle
surface characteristics.  The electron microscope
can size 0.002μ diameter particles.

Mobility--either thermal, electrical, magnetic or
concentration gradients can produce particle size
separation for particles 0.01μ diameter and larger.

Piezoelectric--some crystals can serve as micro-
balances to detect the mass of deposited particles
down to 0.005μg.

Sedimentation--particle mass and medium drag can be
used to size 1 to 100μ particles.

Sieving--coarse particle separation can be affected
as a function of particle size and shape.

Other physical properties, such as rates of
evaporation and cooling and vapor pressures, are
also used as sizing methods.

7.1.2   SIZE DISTRIBUTIONS

   Particulate matter existing in the atmosphere
will usually exhibit a normal size distribution

frequency. It is possible, by various graphical
procedures, to represent this size distribution as
straight lines on a probability plot. This is
convenient in that it enables visual inspection of
data which can help ascertain the validity of
particulate sampling data and procedures. It is
perhaps easiest to describe the graphical procedure
using an example. Assume that the first two columns
of Table 7.2 represent a normal atmospheric dust
and the data has been obtained by some physical
measurement procedure.

The data of Table 7.2 can be relisted to show
cumulative size distribution. This is presented
in Table 7.3.

TABLE 7.2    A TYPICAL ATMOSPHERIC DUST FREQUENCY
            DISTRIBUTION

| *Mass Frequency (n), %* | *Diameter Size Range, μ* | *Range Average Diameter, μ* |
|---|---|---|
| 1 | 1.0 | |
| 1 | 1.0-1.2 ⎫ | |
| 3 | 1.2-1.6 ⎬ | 1 |
| 4.5 | 1.6-2.0 ⎭ | |
| 27.5 | 2-4 | 3 |
| 23 | 4-6 | 5 |
| 15 | 6-8 | 7 |
| 9 | 8-10 | 9 |
| 5.6 | 10-12 | 11 |
| 3.4 | 12-14 | 13 |
| 2.2 | 14-16 | 15 |
| 0.8 | 16-18 | 17 |
| 4 | >18 | |

TABLE 7.3    REARRANGING DATA OF TABLE 7.2

| *Cumulative % by Wt.* | *Less than Diameter, μ* |
|---|---|
| 96.0 | 18 |
| 95.2 | 16 |
| 93.0 | 14 |
| 89.6 | 12 |
| 84.0 | 10 |
| 75.0 | 8 |
| 60.0 | 6 |
| 37.0 | 4 |
| 9.5 | 2 |
| 5.0 | 1.6 |
| 2.0 | 1.2 |
| 1.0 | 1.0 |

(As we proceed, keep in mind that this data could
have been presented by number, area, or any other
measurement system such as indicated in Table 7.1
instead of by mass as these data are.  Note the
common engineering fallacy of considering mass and
weight as being synonymous.  Weight is a unit of
force which can be obtained by multiplying the mass
by the local gravitational acceleration and dividing
this by the gravitational acceleration constant $(g_c)$
which is

$$32.174 \ \frac{ft \ lb_m}{sec^2 lb_f} \ .)$$

Figure 7.1 is a typical S-type distribution
curve obtained by plotting cumulative percent
undersize by weight versus particle diameter in
microns.  Figure 7.2 is a curve obtained by
plotting the slope of the curve of Figure 7.1
versus particle diameter (which is essentially the
same as plotting percent frequency versus an
average diameter for each size range if the size
ranges are chosen small enough).  Note that the
ordinate of the plots could be percent oversize
just as well as percent undersize with the
correspondingly inverted curve.  Figure 7.3 is the
same plot as Figure 7.2 but using log coordinates
for the abscissa instead of the cartesian
coordinates.  This produces the Gaussian type of
symmetrical distribution with most atmospheric
dusts.

The mode (value with greatest frequency for
this data) occurs at approximately 3 microns
diameter as can be seen from Figure 7.2.  The
median (middle value) is readily observed from
Figure 7.1 as occurring at approximately a diameter
of 5 microns.  This is also shown at the right of
the mode in the positive skewed curve of Figure 7.2
which is typical for normal atmospheric dust.

The fact that a Gaussian type normal frequency
curve is obtained in Figure 7.3 indicates that this
data will result in a straight line when plotted on
logarithmic (geometric)-probability coordinates.
Figure 7.4 is a plot of this data on log-probability
paper.  Strictly speaking, the median refers to the
number average and other averages which occur at
50% probability on log-probability plots are
geometric means.  The mean diameters occur at a
probability distribution of 50% and are called $d_{50}$.
When referring to log-probability distribution,
the $d_{50}$ are geometric means and when referring to
Cartesian (arithmetic) probability, the $d_{50}$ are

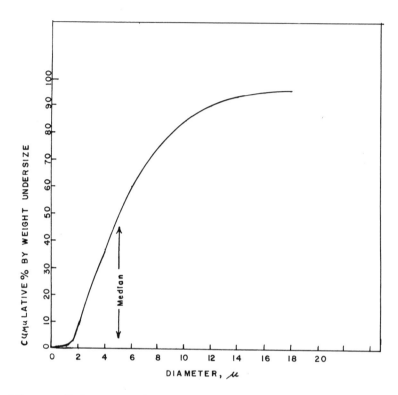

Figure 7.1  A Typical Atmospheric Particle
Distribution; Cartisean (Arithmetic)
Coordinates

arithmetic means.  The geometric mean is obtained
directly from Figure 7.4 or it can be found
mathematically by taking the nth root of the
product of n terms of data:

$$d_{50} = \left(x_1 x_2 x_3 \ldots x_n\right)^{1/n} \qquad (7.1)$$

where x = the individual diameter or area or
volume of the n particles in the
sample

The arithmetic mass mean for this data would occur
at a diameter of 12 microns which is further to the
right on the frequency curves than the median.  The
arithmetic number mean diameter will have a value

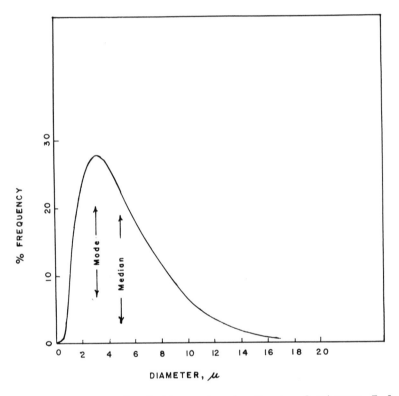

Figure 7.2   Typical Atmospheric Dust of Figure 7.1
             Plotted as Normal Frequency Distribution-
             Positive Skewness on Cartisean Coordinates

closer to the geometric mass mean value (numbers
and masses of spheres are related by $d^3$).   Remember
that the geometric mean is not the same as the
arithmetic mean which is described in Table 7.1.
Means can be considered as moments:   the arithmetic
mean has the property that the algebraic sum of
deviations from it is zero, and for the geometric
mean it is the sum of the products that equal zero.
When using number averages, the two types of averages
will have nearly the same value--although the median
should have the lower value.

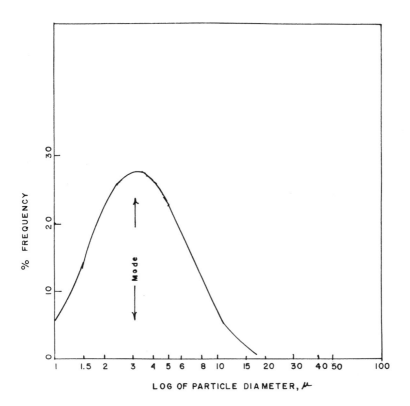

Figure 7.3.  Typical Atmospheric Dust of Figure 7.1
Plotted as Log Normal Frequency
Distribution-Symmetrical Gaussian
Curve

The geometric standard deviation, which is a
measure of the spread of particle diameters from
the mean, can be obtained either directly from
Figure 7.4 or may be obtained mathematically. For
this, use the procedure outlined in Chapter 3, but
use the geometric mean diameter. Standard
deviation ($\sigma$) is obtained most simply from the log-
probability plots either by dividing the diameter
which occurs at a probability distribution of
84.13% by $d_{50}$, or by dividing $d_{50}$ by the diameter
which occurs at a distribution of 15.87%:

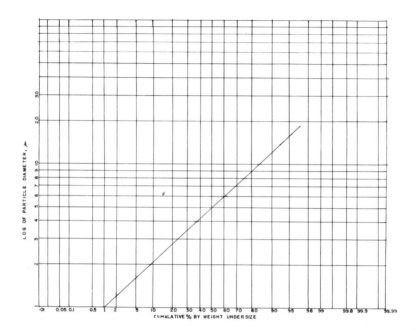

Figure 7.4    Typical Atmospheric Dust of Figure 7.1.
Plotted on Log-Probability Coordinates

$$\sigma = \frac{d_{84.13}}{d_{50}} = \frac{d_{50}}{d_{15.87}} \qquad (7.2)$$

The deviation of the sample data in Figure 7.4 is
2.0 which is a typical value for log normal distribution of atmospheric dust.

Plots of some atmospheric dusts are symmetrical
on Cartesian frequency coordinates (Figure 7.2).
This usually occurs when the mean diameter is large
or when the standard deviation is large.  If this
is the case, then this data should be plotted on
Cartesian-probability paper and not on log-
probability paper.  The arithmetic standard
deviation can be obtained from a Cartesian-
probability plot by subtracting the arithmetic
mean diameter from the diameter whihh occurs at
a probability of 84.13% or by subtracting the
diameter which occurs at 15.87% from the mean
diameter:

$$\sigma = d_{84.13} - d_{50} = d_{50} - d_{15.87} \qquad (7.3)$$

Raw data showing the amount of particulate matter for various size ranges can be plotted on either log-probability or Cartesian-probability paper to help ascertain the validity of the data. Data can be converted from mass measurements to number (or area) percent or vice-versa by using the probability plots.

A suggested procedure for converting is to pick a small interval of cumulative percent at either the high or low percent end of the plot (usually 2% or less). Record the interval and appropriate diameter, then approximately double the interval and use thet result as the interval for the second group... until the *first* 50% of the data are used. Continue taking data from the probability plot, but reverse the procedure for the last 50% (i.e., continuously decrease the interval size). For example: the first group could consist of material equal to or greater than 98% of the material--the average diameter for this 2% of the material is then noted; the second group then could be from 94 to 98 cumulative percent--the average diameter for this 4% of the group is noted; redouble the interval for the next group and note the diameter appropriate to the size range, etc. This results in a listing of frequencies and average group diameters by mass, area, number or whatever the basis of the original plot. The same diameters can be maintained and the number of occurrences in the new base can be calculated according to either the square or cube as indicated by Table 7.1 to give a frequency of occurrence in the new base.

The results obtained by this procedure should be checked by: (1) comparing calculated and graphical mean diameters and standard deviation (if in error, start over using smaller initial interval size); or (2) repeat using smaller interval size and compare the successive results obtained with the previous results. Continue taking smaller intervals until no significant change occurs in the mean diameters and standard deviation between successive checks. It should be noted that conversion of data from mass to area and area to number should result in no significant change of standard deviation. The plots of the same material by these three methods on probability paper result in a parallel family of lines as is shown in Figure 7.5. The mean diameter, however, changes with the largest value being for data plotted by volume, a smaller mean diameter value for data by area and the smallest of the three obtained for data plotted by number.

## 7.1.3   DISTRIBUTION FUNCTIONS

The frequency distribution functions can be plotted using the normal distribution function:

$$\frac{dn}{dd} = f(d) = \frac{1}{\sigma\sqrt{2\pi}} \exp\left[-\frac{(d-d_{50})^2}{2\sigma^2}\right] \tag{7.4}$$

where values of $d_{50}$ and $\sigma$ are arithmetic because we will consider the arithmetic relationship first.

The integral of the normal standard curve with arithmetic frequency becomes:

$$F(Z) = \frac{100}{2\pi}\int_{-\infty}^{Z} \exp\left[-\frac{t^2}{2}\right] dt \tag{7.5}$$

where:   $F(Z)$ = % probability
$Z$ = linear distances along abscissa from center

$$t = \left(\frac{d-d_{50}}{\sigma}\right)$$

The expression t is an arbitrary factor used for simplicity in integration, i.e., an integrating factor. At the center of the graph paper on the abscissa, $Z = 0$ and $F(Z)$ = 50% probability. The following other relationships can be determined:

| Z | $-\infty$ | -2.0 | -1.0 | -0.26 | 0 | +0.26 | +1.0 | +2.0 | $+\infty$ |
|---|---|---|---|---|---|---|---|---|---|
| F(Z) | 0 | 2.3 | 15.87 | 40.0 | 50 | 60.0 | 84.13 | 97.7 | 100 |

A plot of these values of F(Z) at distances indicated by Z on the abscissa of Cartesian co-ordinates produces the arithmetic probability coordinates.

The log normal probability distribution function is:

$$\frac{dn}{dd} = f(d) = \frac{1}{d\sigma\sqrt{2\pi}} \exp\left[-\frac{(\ell nd-\ell nd_{50})^2}{2\sigma^2}\right] \tag{7.6}$$

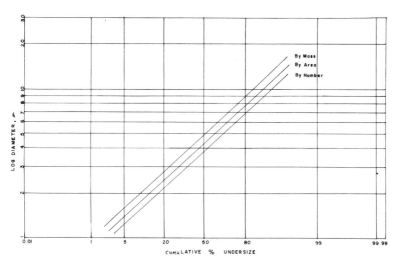

Figure 7.5    Family of Curves Comparing Probability
by Mass, Area and Number

where $d_{50}$ and $\sigma$ are geometric because of the
logarithmic relationship.
The integral expression for the log normal
standard curve is:

$$F(Z) = \frac{100}{\sigma\ \sqrt{2\pi}} \int_{-\infty}^{Z} \exp\left[-\frac{t^2}{2\sigma^2}\right]dt \qquad (7.7)$$

where:    $t = \ln \dfrac{d}{d_{50}}$

Similar substitution of Z values for distances from
the center (Z = 0) gives the same probability scale
as before, but this time developed using log-
probability type data.

7.1.4    MATHEMATICAL-GRAPHICAL SIZING SUMMARY

It should be obvious that the simplest method
of fairly accurately describing an atmospheric
dust is to present the mean diameter and the
standard deviation. In addition, it is necessary
to specify whether the mean has been obtained by
number, area, volume, etc. and if it is log normal
or Cartesian normal distribution frequency. A
further qualification would be necessary if it is

not possible to consider the particle as being spherical in shape.

Mathematical sizing requires frequency values (n) for each small range of particle diameters. The arithmetic mean diameter (d) is then:

$$\bar{d} = \underset{By\ Number}{\left(\frac{\Sigma nd}{\Sigma n}\right)} \quad \underset{By\ Area}{\left(\frac{\Sigma nd^2}{\Sigma n}\right)^{1/2}} \quad \underset{By\ Mass\ (Volume)}{\left(\frac{\Sigma nd^3}{\Sigma n}\right)^{1/3}}$$

The arithmetic standard deviation is found by:

$$\sigma = \left(\frac{\Sigma n(d-d)^2}{\Sigma n}\right)^{1/2} \tag{7.8}$$

for all cases.

Graphical sizing can be carried out as soon as a check has been made to determine whether log frequency or Cartesian frequency distribution produces the most symmetrical distribution curve. (If in doubt, use log distribution.) Convert data to cumulative percent undersize (or oversize) by the following procedure:

$$\% \text{ undersize} = \underset{By\ Number}{\left(\frac{n\ at\ given\ d}{\Sigma n}\right)100} \quad \underset{By\ Area}{\left(\frac{nd^2}{\Sigma nd^2}\right)100} \quad \underset{By\ Mass\ (Volume)}{\left(\frac{nd^3}{\Sigma nd^3}\right)100}$$

Then cumulative % undersize = $\Sigma$ % undersize at the given d's. Plot log of diameter versus cumulative % undersize on log-probability paper (if the log frequency plot gives the most symmetrical curve). From this plot, the geometric standard deviation is obtained using the geometric mean diameter ($d_{50}$) and:

$$\sigma = \frac{d_{84.13}}{d_{50}} = \frac{d_{50}}{d_{15.87}} \tag{7.2}$$

If the Cartesian frequency plot gives the most symmetrical curve, then the data should be plotted on Cartesian-probability coordinates. Arithmetic standard deviation is then obtained using the arithmetic mean ($d_{50}$):

$$\sigma = d_{84.13} - d_{50} = d_{50} - d_{15.87}$$

The lines obtained when the data are plotted on probability coordinates frequently have S-shaped ends. Although this is normal, the bulk of the data should produce a nearly straight line. Substantial deviation from this could indicate any of the following: bad data points, the presence of two distinct types of particulates in large quantities, or that something has happened to the sample to create a non-normal type of distribution (e.g.. removal of a fraction of the sample by screening or depositing; thermal expansion and subsequent breakup of some of the particles in the sample; etc).

## 7.1.5   SHAPE VARIATIONS

Although liquid particulate matter usually can be considered to be spherical in shape, solids frequently should not. Aerosol fumes of metals that form by condensation can be highly irregular in shape. In addition to shape irregularity, they can contain air trapped internally which also results in density variations. It has been proven (1) that the density of aerosol flocs are commonly 10-30 times lighter than the normal density of a solid chunk. Hydroscopic aerosols absorb water and the density of these tend to approach the density of water. Deliquescent particles that absorb water and are soluble too, become spherical in shape resulting in new particles that vary from the original particles in both shape and density. Both irregular shapes and the absorption of moisture result in increased light scattering, and atmospheric haze is caused by the presence of these types of particulate matter.

An example of particulate matter classified by shape is given below:

*Spherical*--carbon black, starch, iron oxide fumes,
            PVC and other plastics, glass beads,
            pollens and spores.
*Rectangular*--iron powder, quartz and other minerals
*Splinter*--cement, organic pigments, corundum $Al_2O_3$
*Platelet*--mica, graphite, bronze
*Rodlets*--talc, flour
*Fibrous*--cellulose, textile fibers

In addition to having shape irregularities, aerosols agglomerate at a rate proportional to a second order chemical reaction, as discussed in Section 9.4. This coagulation changes both the final size and physical structure of aerosols.

Section 7.1.1 describes that shape variations could be accounted for by assigning drag or Stokes' equivalent diameters. In addition, we can account for deviation from sphericity by procedures such as that of Pettyjohn (2). This procedure suggests the use of a dimensionless shape factor (K'):

$$K' = 0.843 \log \frac{\Psi}{0.065} \qquad (7.9)$$

where the sphericity factor ($\Psi$):
= 1, sphere
= 0.906, cubical octahedron (8 equal sides - 4 triangles each on top and bottom)
= 0.846, octahedron or rodlet
= 0.806, cube or rectangle
= 0.670, tetrahedron or splinter

The highly irregular platelet and fibrous form of pollutants deviate too greatly from a sperical configuration to be represented by the sphericity factor.

## 7.2  GASES

There are no special classifications to describe physical geometry or shape variations of gases. Chemical properties are discussed briefly in Chapter IV. In addition to chemical properties, gaseous pollutants are frequently classified by odor, color, density and special hazardous property (e.g. poisonous or explosive).

Gaseous pollutants in the ambient air are usually well below the lower explosive limit. Color and density serve to identify some pollutants, but odor is the most distinctive physical feature.

The human olfactory system is highly complex and so far, no completely suitable analytical substitutes are available. The odor perception threshold is the lowest concentration of a substance where its odors can be perceived. Although this occurs at very low concentrations, there is a wide range between various pollutants. For example, ammonia is considered to be highly odorous, yet it takes a concentration of 47 ppm before it can be perceived by smelling; yet, isovaleric acid has a perception threshold of 0.6 ppb, which is almost 100,000 times more dilute (3). An individual's odor perception is not consistent and varies from time to time as well as between individuals. There is therefore no accurate odor classification procedure. Odor measurements and odor dilutions are discussed in the last chapter of this book.

7.3  MIXTURES

It is worthwhile to mention that mixtures of particulate and gaseous pollutants can be classified by an opacity method using the Ringelmann Smoke Chart (4). This chart, as shown in Figure 7.6, represents varying degrees of darkness (each chart has 20% more black area than the preceding chart). This method of classification is an art and not a scientific procedure; therefore, it is mentioned under this section and not under Analytical Procedures. To use this chart, an inspector views smoke for a given time and at given frequencies. He then relates its blackness to one of the Ringelmann Charts. Chart No. 0 means clear or no smoke and Ringelmann No. 5 means totally black. The average percent of smoke density then equals the sum of Ringelmann Numbers X 20 divided by the number of observations. Curently, the preferred methods of obtaining smoke measurements use optical scattering and transmittance. These methods which include telephotometry, photography, photometry, smoke guides and lasers are discussed by Conner (5). Turk (6) presents methods of exhaust smoke odor evaluation using odor (smell) discrimination. Odor threshold concentrations for various percentages of observers are listed by Leonarda (7) and others.

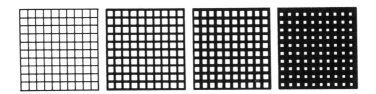

Figure 7.6  Ringelmann Smoke Chart

QUESTIONS FOR DISCUSSION

1. A laboratory analysis has yielded data concerning the concentration of iron oxide in an industrial atmosphere. The data were obtained by noting weight increase on filters having various pore sizes. How would you summarize this data for the most concise and accurate presentation?
2. What is the difference between a median, an arithmetic mean, and a geometric mean? Give examples.
3. What is the difference between arithmetic standard deviation and geometric standard deviation?
4. How would you characterize a gaseous pollutant released into the atmosphere?
5. Discuss the possible physical characteristics of aluminum oxide fume both at a short time after release and after longer times from release.
6. A particulate sample is known to have a mean diameter by number of 9.5 microns and a standard deviation of 1.8. Could this material be considered to be a normal type of atmospheric dust? Are these values arithmetic, geometric or something else?

PROBLEMS

7.1 A sample of particulate matter released from an industrial heating system stack has the following distribution (at 1 grain/ft ) in cumulative weight percent undersize:   3% at one micron diameter, 20% at 2 micron, 42% at 3 micron, 60% at 4 micron, 72% at 5 micron, 81% at 6 micron, 87% at 7 micron, 91% at 8 micron, 94% at 9 micron and 95% at 10 micron.
    a.  Plot this data as both Cartesian and logarithmic frequency distribution curves.
    b.  Plot the data on the appropriate probability coordinates (then plot it on the other probability coordinates for comparison--note this fact on the plot).
    c.  What is the value of the load, the mean and the standard deviation?
7.2 Ball milled clinker dust is reported to have a standard deviation of 3.6 and a mean diameter by number of 15 microns.
    a.  Find the expected frequency distribution for this material.
    b.  Calculate the mean diameter by area and by mass.
7.3 Convert the data of 7.1 to a probability plot showing distribution by number. What is the value of the median?

<u>7.4</u>  A particulate sample is reported to have a mean diameter of 3 microns and a standard deviation of 1.8.  Plot this data on appropriate probability coordinates and determine what percentage of the sample is less than 5 microns, 10 microns and 15 microns.
<u>7.5</u>  Use a magnifying glass or microscope to classify some dust sample.  Report all the information as completely and concisely as possible.
<u>7.6</u>  Repeat probem 7.1 using the following size distribution data:  3% at 1µ, 8% at 5µ, 15% at 10µ, 24% at 20µ, 41% at 40µ, 58% at 60µ, 69% at 70µ and 92% at 80µ.  These data represent fly ash from a large water tube boiler (power generator) using a chain grate fired stoker.

REFERENCES

1.  Whythlaw-Gray, R. and Paterson, H.H., "Smoke," Edward Arnold & Co., London (1932)
2.  Pettyjohn, E.S. and Christianson, E.B., "Effects of Particle Shape on Free-Settling Rates of Isomertric Particles," CEP, Vol. 44, No. 2, pp. 157-172 (1948)
3.  Hanna, G.F., "Odor Measurement Methods," AIChE Odor Control Symposium Feb. 15-19 (1970)
4.  Kudlich, R., "Ringelmann's Smoke Chart," as revised by L.R. Burdick, W.S. Bureau of Mines, Information Circular 7718 (1955)
5.  Conner, W.D., and J.R. Hodkinson, "Optical Properties and Visual Effects of Smoke-Stack Plumes," U.S. Department of Health, Education and Welfare Publication Number 999-AP-30 (1967)
6.  Turk, Amos, "Selection and Training of Judges for Sensory Evaluation of the Intensity and Character of Diesel Exhaust Odors," U.S. Dept. or Health, Education and Welfare Publication Number 999-AP-32 (1967)
7.  Leonarda, G., D. Kendall and N. Barnard, "Research on Chemical Odors--Odor Thresholds for 53 Commercial Chemicals", TAPCA, Vol. 19, No. 2, pp. 91-95 (1969)

# CHAPTER VIII
# COMBUSTION AND RELATED
# POLLUTANTS DISPOSAL

Pollutants that are generated by combustion processes and industrial operations usually require ventilating systems and stacks to remove these wastes from the immediate vicinity in which they are produced in order to make it possible for humans to function in the area. Industrial hygiene standards specify design and operating criteria for ventilating facilities to protect the health of the industrial workers.

Stacks *can* effectively remove gaseous emissions from the immediate area. In addition, stacks aid in enabling the off-gases to be diluted and aid in drawing needed air into combustion operations. The disadvantage arises in the fact that the released stack gases create problems downwind if the emissions are excessive or if other contributing factors are unfavorable. (Plume dispersion is discussed in Chapter III).

This chapter presents general requirements for ventilation of localized atmospheres and includes some design procedures for stacks. (See also Section 11.8 of Control Equipment.) Most combustion gases are released through stacks, so for this reason, both combustion theory and stacks are discussed in this chapter. Combustion of coal, oil, and gas as sources of energy are specifically included because of the significance of their pollution problems. Combustion is oxidation, so both thermal and catalytic oxidation control procedures and devices are included in this chapter.

## 8.1 STACKS

Stacks are the most common method for disposing of gaseous waste products into the atmosphere. In general, the higher the stack, the further away the

pollution effects will be noted, although turbulent
wakes of buildings and downwash can occur in the
vicinity of the stacks, causing local fumigation as
discussed in Chaper III. Downwind ground level
concentration varies inversely as the square of the
stack height. Stacks range in capacity from
extremely small vents to systems that handle
millions of cubic feet of gases per minute. Air
pollution released in small quantities from stacks
can 1) be blown away by the wind and diffused with
the atmosphere until acceptable concentrations are
attained; 2) physically fall out to the ground; or
3) change form by chemically reacting.

It is possible to reduce the pollution from
combustion or process off-gases by changing the
method of operation or by changing the fuel, but
this procedure does not always produce a net
reduction in air pollution. As an alternative, control
devices can remove pollutants from the waste gas
stream before they reach the stack. For example,
settling chambers, filters, heat exchangers,
scrubbers and precipitators can be built into the
stack assemblies. Stacks will continue to be used,
but should be understood and used to the best
advantage.

8.1.1. STACK DRAFT

It is desirable to have exit stack gas velocities
of at least 60 ft./sec. for minimizing air pollution
effects. This velocity helps create added plume rise
which results in a higher effective stack height.
During low level inversions, high exit stack
velocities are needed to help the gases escape by
literally enabling the gas stream to "punch" through
the inversion cap. It should be pointed out that
natural draft cannot create stack exit velocities
much over 10 ft./sec. Booster fans are required
for the high velocities, but natural draft still
assists.

Stacks produce natural draft by the bouyant lift
of the warm air. Natural stack draft is a function
of exit gas temperature, local atmospheric
temperature, stack height (atmospheric pressure and
density differences at the top and bottom of the
stack are directly proportional to stack height),
and wind velocity over the top of the stack.
Theoretical draft at the base of the stack is
estimated as follows, neglecting wind velocity:

$$D_r = \left(\frac{H'}{5.2}\right)(\rho_{ca} - \rho_{HG}) \qquad (8.1)$$

where:   $D_r$ = draft, inches of water

H' = stack height, feet

$\rho_{ca}$ = density of cold air outside stack, lb./ft.$^3$ (=0.0743 lb./ft.$^3$ at 60°F and 1 atm)

$\rho_{HG}$ = density of hot gas inside stack, lb./ft.$^3$

The net effective draft (D'$_r$) is obtained by subtracting the kinetic energy (KE) and friction losses (F') from the theoretical draft equation:

$$D'_r = D_r - KE - F' \qquad (8.2)$$

Both KE and F' are in inches of water. The kinetic energy is equal to:

$$KE = \left(\frac{12\rho_{HG}}{\rho_{H_2O}}\right)\left(\frac{v^2}{2g_c}\right) \qquad (8.3)$$

where:   $\rho_{H_2O}$ = density of water, lb./ft.$^3$

v = stack exit velocity, ft./sec.

$g_c$ = gravitational constant = $32.174 \dfrac{ft. \; lb._m}{sec.^2 lb._f}$

Friction losses in a stack vary directly with stack height and inversely with the fifth power of stack diameter (D$^{-5}$). The friction losses can be calculated by:

$$F' = \left(\frac{12\rho_{HG}}{\rho_{H_2O}}\right) F \qquad (8.4)$$

The friction loss in feet of fluid flowing (F) can be calculated by the Fanning equation which is applicable for turbulent or viscous flow systems in which the fluid density, viscosity and linear velocity are constant:

$$F = 4f\left(\frac{v^2}{2g_c}\right)\left(\frac{L}{D}\right) \qquad (8.5)$$

The length (L) is the sum of stack height and duct length, but additional equivalent lengths which result due to the presence of bends, etc., should be added (an approximation would be to add 50 D to L). The friction factor (f) is based on experimental data and is a function of the Reynolds' number ($N_{Re}$) and relative roughness of the pipe. Exit velocities of 60 ft./sec. can result in $N_{Re}$ > $10^6$, indicating turbulent gas flow in the stack. The friction factor for turbulent flow and a smooth stack can be estimated by:

$$f = \frac{0.04}{N_{Re}^{0.16}}$$

### 8.1.2.  STACK DESIGN CONDITIONS

Normally, the temperature drop of gases in the stack is approximately $1^\circ F$ per foot of height. If this is sufficient to cause significant variation in the gas density, viscosity and velocity, it may be necessary to make the friction and kinetic energy calculations of Section 8.1.1 over small intervals using appropriate average temperatures for each interval. There may be times when it will be necessary to add friction losses due to entrance and exit constriction and enlargement, respectively, as well as friction losses due to the presence of air pollution control devices or other accessories. An approximation would be to add $3v^2/2g_c$ to the value of F from Eq. (8.5).

A typical stack releasing a combustion type off-gas frequently has the average values listed below:

   Hot gas temperatures = $860-960^\circ R$ $(400-500^\circ F)$
   Atmospheric cold air temp. = $522^\circ R$ $(62^\circ F)$
   Coefficient of friction (f) for rough concrete
      = 0.016
   Hot gas density at $0^\circ F$ and 1 atm. = 0.09 lb./ft.
   Sea level pressure = 29.92 in. mercury
   Exit gas velocity = 60 ft./sec. (minimum)

The required chimney height above the inlet when natural draft is used may be approximated by:

$$H' = 190 \ D'_r \qquad (8.7)$$

Evaluation of the previous equations shows that
*forced* draft is required to obtain velocities of
60 ft./sec.  If air pollution control is adequately
perfected so that emissions are negligible, exit
velocities (and stack heights) may be much lower
than now recommended.  The maximum stack diameter
is calculated by taking advantage of the information
that the minimum stack exit velocity should be
60 ft./sec.

If stack gases are cooled by air pollution
control devices, such as wet scrubbers, the
effective height of the stack is drastically
reduced.  It is occasionally necessary to reheat
and/or use booster fans to obtain the necessary
stack velocity.  Under no circumstances should the
gas be forcefully released at a temperature lower
than the ambient air temperature.

Recent developments in structural materials
make possible short stacks of lightweight materials.
Frequently, steel or coated steel stacks (supported
by guide wires) are used for low temperature stacks
up to 100-150 feet in height.  Radial brick stacks
are often the most economical stacks for hot gases
up to a height of about 175 feet.  Beyond this,
reinforced concrete is normally used.  The height
limitations on stacks is a function of both ground
strength and stack strength.  Corrosive gases need
special stack linings; though control devices will
remove these gases before they reach the stack.  An
example of one new type coated stack is the
polyvinyl steel (PVS) which utilizes polyvinyl
chloride commercially bonded to steel tubes.  This
stack is non-clogging, light weight and economical.
A disadvantage is that stack temperatures are
limited to a maximum of about 450°F.  Beyond this
temperature, the plastic begins to decompose and
forms poisonous hydrochloric acid and/or phosgene.
Other stacks are made of fiber glass, specially
resistant refractory or lined, for example, with
rubber.

## 8.2   COMBUSTION THEORY

Combustion or incineration is normally
considered to be the burning of a volatile material
with air.  This may be to decompose and thereby
dispose of the "fuel" as gaseous vapors, or to
generate energy in the form of heat.  Ideally, it
should be possible to use combustion to decompose
unwanted material as well as to generate heat

during the deocmposition.  Unfortunately, most of the
unwanted material is not in a form to be economically
used as fuel sources for obtaining heat energy.
(See Section 8.5 for refuse combustion.)
     Combustion is a chemical reaction that requires
the contacting of a fuel with oxygen at a temperature
above the kindling temperature.  Both a high degree
of turbulence and adequate oxygen are required to
attain complete combustion.  The stoichiometric air-
fuel ratio which is theoretically required for
complete combustion according to the balancing of
the chemical equation, varies with the fuel.  It
was shown in Capter VI that the air-fuel ratio for
burning gasoline is approximately 15 pounds of air
per pound of fuel.  the ratio for any other fuel
can be calculated in a similar manner.
     Excess air is the ammount of air added to a
combustion operation beyond that required stoichiometrically by the chemical equation.  For example,
100% excess air is twice the amount of air stoichiometrically required for complete combustion.  In
practice, values of about 12-20% excess air are
common for combustion type operations, although the
current trend is to use less (2-3%) excess air to
reduce formation of nitric oxide.
     The desired gaseous products of complete combus-
tion are--all hydrogen in the form of water and all
carbon in the form of carbon dioxide.  In addition
to this, any other materials present can be oxidized
to their normally highest oxidation state.  For
example, sulfur is oxidized to sulfur dioxide and
metals are oxidized to their oxides.  Common products
of incomplete combustion are carbon (soot), CO,
hydrocarbons of various types and hydrogen chloride.
Remember that other solid products of combustion
are formed such as ash and clinker.  The ratio of
metals to metal oxides present varies depending on
the amount of excess air and furnace design which
determines the extents of the reducing and oxidizing
zones.
     The nitrogen present in combustion reactions
can come from both the air and the fuel.  Some of
the nitrogen is oxidized, with nitric oxide (NO)
being an undesirable product of combustion.  The NO
formed is a function of flame temperature, reaction
rates, residence time, nitrogen and oxygen
concentrations and quench rate.  Ermenc (1) suggests
the simplified curve shown in Figure 8.1 for
estimating the NO conversion as a function of flame

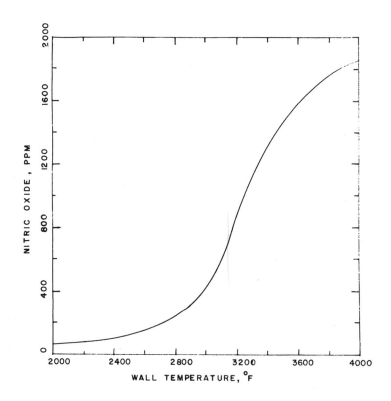

Figure 8.1.  Nitric Oxide Produced During Combustion
as a Function of Temperature in a
Production Furnace

temperature. (Actual combustion flame temperatures
for hydrocarbons range from 1600-3600°F, with
theoretical flame temperatures slightly higher.)
    The equation showing oxidation of nitrogen by
oxygen to form nitric oxide is:

$$N_2 + O_2 \rightleftharpoons 2NO \qquad (8.8)$$

Ermenc showed that at temperatures above 2000°F,
the rate equation is second order and that the rate
of the reverse reaction is much greater than the
rate of the forward reaction. The approximate
values of the rate constant for the forward reaction
at 3200 and 4000°F respectively, are 5 X $10^{-2}$ and
100 atm$^{-1}$sec$^{-1}$; the reverse rate constant at the

same temperatures is 85 and 5 X $10^4$ $atm^{-1}sec^{-1}$. Considering both temperature and contact time in typical combustion operations, the forward reaction controls at temperatures that are elevated but less than $3200°F$ (because of the time available) and the reverse reaction controls at temperatures above $3200°F$ (because of the shorter time and high rate of the reverse reaction). Slow cooling of combustion gases at temperatures above $3200°F$ would cause the NO to revert to nitrogen and oxygen, while rapid cooling to $3200°F$ fixes the higher NO content. Pollution control, therefore, requires the slowest possible high temperature region cooling rate. Processes that chill very hot gases by passing them into pollution collectors actually increase NO pollution. (Autos have "chilled" exhaust which helps contribute to the high NO emission.)

As excess air and turbulence in a combustion chamber are increased, more products of complete combustion are obtained. As excess air is increased, combustion efficiency increases to a maximum and then decreases with further increase of excess air. This results because of the cooling effect that occurs when some of the heat released from combustion is required to heat the incoming excess air to the combustion temperature. When more heat is lost to the stack gases than is gained by improved combustion due to increasing excess air, then maximum overall thermal efficiency must decrease. Up to 5% of the fuel is frequently sacrificed and not burned in order to obtain a higher maximum overall efficiency. Low excess air helps reduce the amount of NO pollution formed and keeps the exhaust blower capacity (costs) to a minimum. There are times, however, when large quantities of excess air (e.g. 200%) can be used to reduce the NO emissions because of both 1) the resulting lower flame temperature and 2) the reduced contact time between $N_2$ and $O_2$ at elevated temperatures.

It is possible to obtain the highest amount of intimate contact between fuel and air when the fuel is gaseous. Figure 8.2 shows the typical flue gas analysis for the combustion of natural gas in air at stoichiometric (0% excess) air and at greater than stoichiometric amounts of excess air. Even with this most easily mixed fuel, actual flue gas analyses will vary depending on burner configuration and temperature. Figure 8.3 shows schematically how the composition of the individual components in flue gas varies with changes in amounts of air

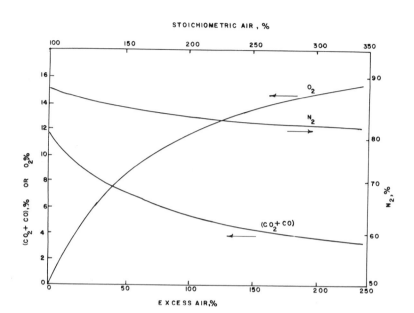

Figure 8.2.  Typical Flue Gas Analysis for
          Combustion of Natural Gas in Air

during combustion. The same degree of mixing is
assumed for all amounts of air. Note that although
the total carbon dioxide is increased with increasing
air, the concentration of carbon dioxide decreases
because of the dilution effect. Thermal efficiency
is also indicated schematically on this diagram.
    So far, we have seen that NO emissions can be
reduced by: reducing the quench rate when the
gases are above 3200°F; decreasing the $O_2$
concentration by reducing the amount of available
air; and by decreasing the flame temperature and
contact time by using large amounts (e.g. 200%) of
excess air. Fuel type also affects NO formation
because the flame temperatures attainable and
radiant heat loss rates differ with fuel. In
general, for medium and small combustion facilities,
the NO emissions are *greatest from coal, less from
oil* and *least from gas.* In large installations,
the reverse is usually true.
    Other techniques used to reduce NO formation
are two-stage combustion, flue gas recirculation
and water injection. Two-stage combustion utilizes

a minimum amount of oxygen in the first stage, gas
cooling between the two stages, then the balance of
the air is used in the final stage. This gives
lower gas temperatures and lower $O_2$ and $N_2$ contact
time. Flue gas recirculation mainly serves to
lower the peak flame temperatures.

Injection of water or low temperature steam
keeps the flame temperatures low and dilutes the
concentration of $O_2$ but also reduces thermal
efficiency. (This quenching should not be done if
the gases are above $3200^\circ F$.) An added advantage
of water injection is that by use of a shift
reaction catalyst, the water is reduced to $H_2$ by
the CO present:

$$H_2O + CO \xrightarrow{\text{shift catalyst}} H_2 + CO_2 \qquad (8.9)$$

This provides added combustion heat, reduces the CO
pollution concentration and forms $H_2$ which reacts

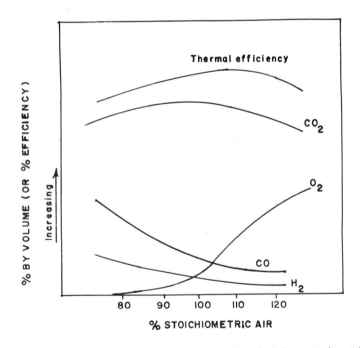

Figure 8.3.   Gas Analyses vs% Stoichiometric Air
(Assuming a constant turbulence which
is not complete mixing--not to scale)

with the $SO_2$ pollution to form S or $H_2S$. Best
results for these reactions occur when the
concentration of CO is 1%, the temperature is
6000-950°F, contact times are 0.2 sec. and using a
copper on alumina catalyst.

A method of eliminating NO formation entirely is
to use fuel with no nitrogen and pure oxygen. There
are obvious problems to this, but it can and is being
done. Studies are also being performed that use CO
and copper, silver, or gold on alumina catalysts to
reduce NO to $N_2$ while forming $CO_2$ from the CO. The
same system forms S or COS from $SO_2$.

## 8.3   COAL COMBUSTION

### 8.3.1   COAL FUEL

Coal, a fossil fuel, is used to produce most of
the electric power in the U.S. (see Chapter II). The
two types of coal are bituminous (soft coal) and
anthracite (hard coal). The chemical composition
of coal varies from one seam to another, and in
addition, there are significant chemical and
physical variations between the bituminous and
anthracite types of coal. Bituminous coal usually
contains more volatile matter than anthracite coal
--it may average 20% as compared to 6% for
anthracite. Anthracite usually has more ash than
bituminous and may have an average of 15% compared
to 10% for bituminous. The sulfur content is lowest
in the anthracite coal and usually amounts to less
than 1%, whereas it may be as high as 4% in
bituminous coal. The heating value of good coal is
over 12,000 BTU/lb. Chemical analyses of a typical
Pennsylvania bituminous type coal on an air dried
basis is given by both a "proximate" and the
ultimate analysis in Table 8.1.

TABLE **8.1**   TWO TYPES OF COAL ANALYSES--ON AN AIR
DRIED BASIS (FOR ONE BITUMINOUS COAL
SAMPLE)

| | *Proximate, %* | *Ultimate, %* | |
|---|---|---|---|
| moisture | 3.2 | Carbon | 79.90 |
| volatile combustible | | Hydrogen | 4.85 |
| matter | 21.0 | Sulfur | 0.69 |
| fixed carbon | 69.3 | Nitrogen | 1.30 |
| ash | 6.5 | Ash | 6.50 |
| TOTAL | 100.0 | Oxygen (by | |
| | | difference) | 6.76 |
| | | TOTAL | 100.00 |

Coal ash contains approximately 50% silica.  It also contains approximately 1/3 alumina and minor amounts of the oxides of iron, calcium, magnesium, titanium, sodium and potassium.  Ash usually also contains sulfur as sulfates, which varies from 0.1 to over 10%.  Clinkers, which are formed mainly from the ash in coal, are fused chunks of metal oxides.  The ratio of metal to metal oxides varies depending on the extent of the boiler reducing zone.

The combustion theory section (8.2) indicates that not only is adequate oxygen required for good combustion, but that it is necessary to intimately contact the oxygen with the molecules of the fuel to be burned at an elevated temperature.  To improve combustion, coal is mechanically broken down into various size grades of fuel.  The coal directly from the mines usually is greater than four inches in average diameter.  There are at least twelve different grades by size which are produced by mechanically cracking or pulverizing the coal.  The smaller the coal is crushed, the greater the surface area exposed and available for combustion; however, the smaller the coal size, the greater the possibility that the coal will be blown out of the furnace by entrainment in the combustion gases.

As the carbon in coal is burned, it can go through three separate chemical reactions.  In the oxygen-low regions, carbon is burned to form carbon monoxide while releasing heat:

$$C + \tfrac{1}{2}O_2 \longrightarrow CO + 52,090 \text{ BTU/lb mole} \qquad (8.10)$$

In regions where there is more oxygen, the carbon can be converted directly to carbon dioxide, and any carbon monoxide available can also be converted to carbon dioxide:

$$C + O_2 \longrightarrow CO_2 + 173,720 \text{ BTU/lb mole} \qquad (8.11)$$

$$CO + \tfrac{1}{2}O_2 \longrightarrow CO_2 + 121,630 \text{ BTU/lb mole} \qquad (8.12)$$

It is apparent, from the amount of heat released in converting to carbon dioxide, that it is most desirable to have low amounts of carbon monoxide in the product gases.  Coal boilers should be designed for maximum air turbulence to eliminate reduction zones and to keep the carbon dioxide from recontacting the coal in the fire box so that the $CO_2$ is not reduced to carbon monoxide with the loss of heat:

$$CO_2 + C \longrightarrow 2CO \quad -69,540 \text{ BTU/lb mole} \quad (8.13)$$

8.3.2 COAL BURNERS

There are three basic types of coal firing: cross-feed; over-feed; and under-feed. The cross-feed bed is one in which the fuel moves at right angles to the air. This appears to be the most desirable for automatic power stations and should result in the least amount of carbon dioxide being reduced to carbon monoxide. The second type is the over-feed bed which has counter-current air fuel movement so that the hot gases heat the incoming coal fuel. A typical example of this type is the hand fired furnace installations. It has the disadvantage in that carbon dioxide becomes reduced by the carbon with the subsequent heat loss. The underfed bed has co-current air and fuel movement and is less common.

8.3.3 COMBUSTION PRODUCTS TO STACK

The combustion theory section pointed out that oxidation of any of the components in the fuel is possible. In coal burning, material that fuses together and remains with the ash does not become airborne in the exhaust gas. As not all the fuel enters the gas stream, it is necessary to consider only the part of the fuel which becomes airborne. Factors which cause increased particulate emission rates are: increasing gas velocity, burning rate and boiler efficiency; and decreasing particle size and density. Fly ash may contain up to 10% combustible material (carbon, carbon compounds, sulfur, etc.). The bulk of the fly ash consists of metal oxides and is whitish-gray in color. As would be anticipated from the discussion of particulates in Chapter VII, fly ash in the atmosphere has a density less than the average value of compact fly ash. The specific gravity of atmospheric fly ash is reported to be 1.0 or slightly less (2) whereas the average bulk density value for fly ash is approximately three times that amount.

It is indicated (3) that typical fly ash from stoker fired boilers has a frequency distribution that by weight would result in an average diameter of 100 microns and a geometric standard deviation of 4.5. (Note that these values result from a

composite of data and, for that reason, are
represented as a log-normal distribution. These
$d_{50}$ and $\sigma$ values indicate arithmetic distribution
and a  artisean-probability plot would be used if
these data were for a single particulate sample.)
When using pulverized coal as fuel, the fly ash
size distribution reported by the same reference
has an average diameter of 20 microns by weight
and a geometric standard deviation of 4.0.
Particulate matter in boiler gases from coal
combustion expressed as dust burden range from 0.5
to 1.0 X $10^{-4}$ lb/ft$^3$ at normal conditions.
   Section 8.2 tells us that when we attempt to
obtain complete combustion during the burning of
coal, we can also oxidize any volatile matter that
exists in the fuels. Sulfur in coal is present
as both elemental sulfur and sulfur compounds
including hydrogen sulfide. Data show that anywhere
from 65-95% of the sulfur in the fuel is converted
to sulfur dioxide during combustion. It is common
practice in air pollution calculations to consider
that *all* of the sulfur in the coal is oxidized
to sulfur dioxide. The small amount of sulfur not
released as $SO_2$, $SO_3$, or $H_2SO_4$ mist leaves the system
either in the fly ash or with the ashes.
   The nitrogen contained in the air portion of the
fuel and in the coal can produce nitrogen oxides
at a rate approximated by the empirical equation
of Woolrich (4):

$$NO_x \, lb/hr = \left(\frac{BTU/hr \text{ heat released from coal}}{3.8 \text{ X } 10^6}\right)^{1.18}$$

$$(8.14)$$

   where:   $NO_x$ = nitrogen oxides as $NO_2$

This equation does not include any variation for
flame temperature and residence time, however, it
can be used as an initial estimate.
   Smith (3) suggests that it is possible to
estimate the amounts of other pollutants in the
gaseous emission from the combustion of coal in
power plants by making a material balance knowing
that the following amounts of gases are released
for approximately every $10^6$ BTU of heat produced:

$$0.02 \text{ lb CO}$$
$$0.007 \text{ lb Hydrocarbons}$$
$$0.0002 \text{ lb Formaldehyde}$$
$$0.08-0.3 \text{ lb HCl}$$

The recommended amount of excess air for coal fired boilers is that value which results in 3-5% $O_2$ in the stack gas. This varies with boiler design and type of coal but amounts to approximately 15-20% excess air. This amount of air results in maximum thermal efficiency and a ghigh percent conversion of the fuel, but it also results in high nitric oxide production. For areas where NO is a problem the current trend is to use only 2-5% excess air.

## 8.4 OIL COMBUSTION

### 8.4.1 OIL FUEL

Fuel oil is a fossil fuel produced as described in Chapter VI from crude oil, coal or other organic starting materials. There are essentially eight grades of oils which can be used for fuel oils: kerosene, diesel oil and the six fuel oils (grades 1 through 6--with diesel oil being essentially the same as number two fuel oil). Each of these become progressively heavier in density and contain carbon compounds with progressively longer chains. In a distillation operation, the kerosene oil would boil at a lower temperature and be removed first and the grade six fuel oil would be one of the last oils to be removed. The common name for grade 6 is Bunker C. This heavier fuel oil was used in steam boilers rated at 34,500 lb steam/hr or greater, although the current trend for the large boilers is to use lighter oils because of their lower sulfur content. Grades 1 and 2 fuel oils are used in smaller commercial oil boilers and for residential heating.

Sulfur content in the fuel oils is of extreme importance because of the potential air pollution problem. Maximum sulfur content in commercial grades of fuel oils and approximate densities and heating values are shown below:

|  | % S | lb/gal. | BTU/gal. |
|---|---|---|---|
| Kerosene | 0.037 | 6.8 | 139,400 |
| Diesel Oil | 0.41 | 7.0 | 141,000 |
| Grade 1 | 0.05 | 7.0 | 137,300 |
| Grade 2 | 1.0 | 7.3 | 141,900 |
| Grade 3 | no limit | 7.4 | 143,400 |
| Grade 4 | no limit | 7.5 | 145,000 |
| Grade 5 | no limit | 7.7 | 148,500 |
| Grade 6 | no limit | 8.0 | 151,300 |

Using kerosene as an example, the following balanced stoichiometric equation shows the complete combustion of kerosene with air (the molecular formula of kerosene is an average molecular formula):

$$C_{12}H_{26} + 18\tfrac{1}{2}O_2 + 69\tfrac{1}{2}N_2 \longrightarrow 12CO_2 + 13H_2O + 69\tfrac{1}{2}N_2$$

$$(8.15)$$

The heating value of kerosene is 20,000-21,000 BTU/lb. If 12% excess (x's) air is used, then the air-fuel ratio (AF) can be calculated by:

$$\text{AF ratio at } 12\% \text{ x's air} =$$

$$(1.12)(18\tfrac{1}{2})\left(\frac{100 \text{ lb air}}{21 \text{ lb } O_2}\right)\left(\frac{28.9 \text{ lb air}}{\text{lb mole}}\right)$$

$$\left(\frac{\text{lb mole}}{170 \text{ lb fuel}}\right) = 16.8$$

Like coal, fuel oils vary in composition depending upon their geological origin. As a result, there are various amounts of minor components present in oil. The heaviest grade of oil can contain up to a maximum of 0.3% by weight ash. The trace minerals, which are present in oil, are essentially the same as those present in coal.

## 8.4.2   OIL BURNERS

There are two basic types of oil burners, tangentially and horizontally fired. Oil is introduced at the circumference of the firebox in tangentially fired boilers and imparts a turbulent motion to the system. Horizontally fired boilers have fuel injected at the sides directed toward the center. This causes the hot combustion gases to concentrate near the center of the unit. As a result, horizontally fired boilers have gases in the center at higher temperatures for longer periods of time which results in the production of approximately twice as much nitric oxide in the horizontally fired units as in the **tangentially fired boilers.**

Good vaporization of the fuel is a necessity to assure good mixing of the oil with the air for proper combustion and to prevent excessive smoking. It is possible to heat the oils until they evaporate, however, this procedure is too slow for commercial use. In lieu of vaporizing the fuel, most commercial boilers utilize atomization procedures because good atomization is essentially equivalent to vaporization. Atomization can be achieved by injecting the oil through spray nozzles, by breaking up streams of liquid oil with high moving jets of air or steam, or by mechanically splashing or spraying the oil as with the use of rotary spray discs (cups).

Atomizing burners that use low pressure air spray nozzles require air supplied by blowers at ½ to 5 psig pressure. The heavy, viscous oils require higher air pressure than the light oils. High pressure burners use 30 to 175 psig air or steam. Both types of atomizing systems handle all grades of oil if the oil is heated to obtain the proper reduced viscosity. Dry steam can serve not only as the atomizing fluid, but it can keep the oil hot. About four pounds of steam are used to atomize each gallon of oil. Air required for atomization can amount up to 200 ft$^3$ per gallon of oil for high pressure systems and up to 1,000 ft$^3$ per gallon for low pressure systems. This amounts to up to 14% and 70% respectively of the total combustion air. The mechanical atomizing burners require high pressure oil at 75-300 psig depending on the oil viscosity. Approximate power requirements to atomize one gallon of oil per hour range from a low of 0.03 hp for mechanical atomizers to 0.4 hp for high pressure air burners (6). (For further information on atomization, see Section 9.7.) Too much excess air in oil burners causes the flame to appear dazzling white. Less air, yielding less NO, produces a yellow flame and, as the flame goes to orange, higher concentrations of particulate matter are produced because there is insufficient air to permit complete combustion.

Fuel oils should be filtered to prevent the clogging of the atomization devices. This is especially true for the heavier fuel oils because they not only contain dirt, but because they are bottoms from distillation facilities and contain tar and sludge. In addition to filtering, the heavier fuel oils must be heated (to approximately 200°F for Bunker C oil) to reduce the viscosity to a point where they can be pumped into the spray nozzles and properly atomized.

## 8.4.3   COMBUSTION PRODUCTS TO STACK

It is necessary to use excess air to help
assure complete combustion. However, as usual,
increased concentration of nitrogen in the presence
of oxygen at high temperatures causes the formation
of undesirable nitric oxide. Woolrich (4) also
provides a formula that can be used to calculate
the amount of nitrogen oxides produced from oil
combustion which is presented here:

$$NO_x\, lb/hr = \left(\frac{x}{248}\right)^{1.18} \qquad (8.16)$$

where:       $x$ = lb oil fired/hr
            $NO_x$ = nitrogen oxides as $NO_2$

Remember that the Woolrich equations do not account
for variation in flame temperature, residence time,
burner design and operating methods. This equation
is only good for horizontally fired units. For
estimating the nitric oxide produced from
tangentially fired units, use half the value
calculated by Equation (8.16). Nitric oxide
emissions can be reduced by: lowering temperature,
two-stage firing, decreasing the load factor (which
decreases temperature), decreasing excess air,
increasing wind box pressure, recycling flue gas,
lowering fuel pressure and frequent cleaning of the
boiler tubes. The optimum amount of excess air for
complete combustion and low NO production results
in approximately 14% $CO_2$ and 3% $O_2$ in the stack gas.
This amounts to about 12% excess air (see Figure 8.4)
which is lower than suggested for coal combustion.
At these conditions, one pound of oil produces
approximately 200 scf of stack gas at 75% thermal
efficiency. Where low NO emissions are required,
the current trend is to use only 2-5% excess air.
Other pollutants released from combustion of
fuel oils are $SO_2$, CO and particulates. Essentially
all of the sulfur in the fuel oil is converted to
$SO_2$. The amount of particulates from fuel oil
combustion is low because the fuel is burned in a
vaporized or highly atomized condition (permitting
intimate contact with air) and the quantity of ash
in the fuel is also low. Emission values for fuel
oil combustion summarized by Smith (5) are presented
in Table 8.2. In addition to those pollutants
listed in Table 8.2, the stack gases contain small
amounts of aldehydes, ketones, hydrocarbnns and
other organics as well as some ammonia and hydrogen.

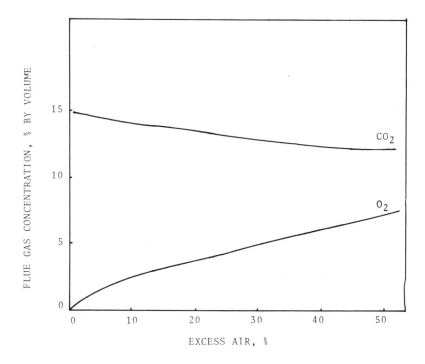

Figure 8.4  Typical Flue Gas Analysis for Combustion
of Oil in Air (Assuming good mixing)

Approximately 2,500 lb oil/hr can produce 1,000
boiler hp.

## 8.5  GAS COMBUSTION

### 8.5.1  GAS FUEL

Gaseous fuels, as discussed in Chapter 6, are
the "cleanest" of the fossil fuels and can be
burned with the least amount of air pollution.
Natural gas is the most commonly used industrial
gas and its use is increasing primarily because of
pollution control requirements.  As a result,
demand is exceeding availability, and the supply
to many gas customers is limited.  Natural gas
consists mostly of methane ($CH_4$) with some ethane
($C_2H_6$), $N_2$ and $CO_2$.  It is obtained naturally from
underground wells.  The heating value ranges from

TABLE 8.2 EMISSIONS FROM FUEL OIL COMBUSTION (5)

| | *Large Sources* *(≥1000 hp)* | | *Small Sources* *(<1000 hp)* | |
|---|---|---|---|---|
| | *ppm in stack gas* | *lb/1000 lb oil* | *ppm in stack gas* | *lb/1000 lb oil* |
| *Emissions* | | | | |
| $NO_x$ as $NO_2$ | | | | |
| horizontal | 470 | 13 | 320 | 9.0 |
| tangential | 210 | 5.8 | | |
| $SO_2$* | $510C_s$ | $19.6C_s$ | $510C_s$ | $19.6C_s$ |
| CO | 0.3 | 0.005 | 15 | 0.25 |
| Particulates | $4X10^{-6}$** | 1 | $4-17X10^{-6}$** | 1.5 |

* Multiply number by weight %
  sulfur $(C_s)$ in fuel oil

** $lb/ft^3$ at SC

950 to 1125 $BTU/ft^3$ and the density is $7.5 \times 10^{-4}$ $g/cm^3$. It is one of the slower burning gases. Table 8.3 summarizes properties of this and some other industrial gases. Liquified petroleum gas (LPG) is mostly propane and butane; and artificial, manufactured or city gas is mixtures of water gas, coal gas and others containing mostly $CH_4$, CO, $H_2$, $CO_2$, $O_2$ and $N_2$. Carbureted (water) gas and producer and coke oven gases are made from coal or coke. Gas fuels contain trace amounts of sulfur (e.g. 0.04% by weight), mostly in the form of hydrogen sulfide and organic compounds such as mercaptans.

8.5.2  GAS BURNERS

Gas burning systems must be designed to mix gas with a desired amount of air, then send the mixture through a properly sized orifice so the flame velocity equals the velocity of the inlet mixture. If the flame velocity is lower than the gas mixture velocity, the flame front will move away from the nozzle until it passes out of the burner and extinguishes because the mixture is no longer in the explosive air-gas range. Too low a mixture velocity causes the flame front to advance to the

TABLE 8.3 APPROXIMATE PROPERTIES OF INDUSTRIAL GASES

| Gas | $CH_4$ | $C_2H_6$ | $CO_2$ | $CO$ | $N_2$ | $H_2$ | SpGr* | Heating Values, Btu/ft³ | STOICHIOMETRIC COMBUSTION | | | Explosive Limits in Air, % | |
|---|---|---|---|---|---|---|---|---|---|---|---|---|---|
| | | | | | | | | | Max. Flame Temp., °F | Max. $CO_2$ Prod., % | Ignition Velocity, ft/sec | Upper | Lower |
| Natural (Mid U.S.) | 96.0 | / | 0.8 | / | 3.2 | / | 0.57 | 967 | 3550 | 11.7 | 1.0 | 4.8 | 13.5 |
| Natural (Pa.) | 67.6 | 31.3 | / | / | 1.1 | / | 0.71 | 1232 | 3600 | 12.3 | 1.0 | 4.8 | 13.5 |
| Artificial | 20 | 1 | 3 | 20 | 12 | 40 | 0.50 | 525 | 3650 | 14 | 2.1 | 5.6 | 34.0 |
| LPG | / | / | / | / | / | / | 1.8 | 3000 | 3660 | 14 | 0.9 | 2.3 | 11.5 |
| Producer | 0.5 | / | 6 | 27 | 55 | 10 | 0.85 | 120 | 3100 | 18 | 0.6 | 18.6 | 73.7 |
| Coke Oven | 25.0 | 1.0 | 2 | 8 | 11 | 48 | 0.40 | 575 | 3600 | 11 | 2.1 | 6.0 | 32.0 |
| Carbureted Blue | 10.0 | 2.0 | 5 | 30 | 15 | 30 | 0.52 | 310 | 3600 | 20.5 | 2.1 | 6.4 | 37.7 |

* Referred to air = 1.0 at SC

burner orifice (spud) where the flame again goes
out. Natural gas, for example, with its low flame
velocity, requires a larger orifice to keep the
mixture velocity low for a given gas volumetric
input rate.

Atmospheric mixers operate with gas at 2-10"
water pressure. The atmospheric air drawn into
the mixer amounts to 40-60% of the stoichiometric
air. The rest of the air comes from the atmosphere
surrounding the flame. Systems employing
atmospheric burners should operate at very low
draft to prevent the flame from being pulled out of
the burner. Where a stack is used, a draft hood is
required to break the draft.

High pressure mixers operate at 1 psig or more
gas pressure and draw in 100% of the necessary
combustion air. The amount of air needed depends
on the gas so the gas pressure varies depending
on how much air it must pull in. For example,
natural gas pressure must be 20 psig while
artificial gas is 10 psig. Proportional and blower
mixing systems use low pressure air (3 psig) or
blowers to draw in or force the gas into the burner.
It is possible to use a mixer valve on these systems
so that it is easy and quick to convert from one
type gas to another.

Excess air in gas fuel systems is not required
to provide turbulence as in the coal and oil systems,
though the burner must be designed to satisfactorily
distribute the heat. Excess air is therefore
minimized to reduce the amount of NO formed and
to reduce heat loss. Figure 8.5 shows heat losses
for a typical boiler resulting from excess air.

### 8.5.3   COMBUSTION PRODUCTS TO STACK

The gaseous combustion products from a properly
operated gas burner should contain essentially $CO_2$
and water with few particulates, sulfur oxides or
incompletely burned organic materials. The amount
of nitrogen oxides produced can be estimated using
either Figure 8.1 or the following modified
Woolrich equation:

$$NO_x lb/hr = [1.65X10^{-4} (cfm)(C)]^{1.18} \quad (8.17)$$

where:  cfm = volumetric gas input rate at STP
        C = wt. % carbon in the gas
        $NO_x$ = nitrogen oxides as $NO_2$

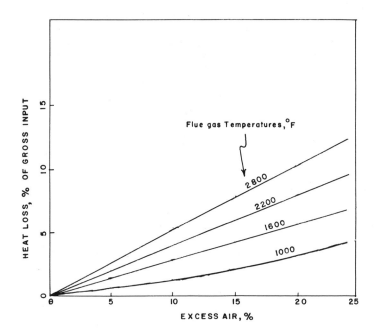

Figure 8.5   Typical Boiler Heat Losses Resulting
             from Excess Air

If the flame temperature cannot be obtained
from Table 8.3, the theoretical flame temperature
can be calculated using the following procedure.
Assuming no heat losses, no side reactions and all
combustion products as gases, the heat of
combustion of the fuel is then available only to
heat up all the products (and any reactants) left
after the combustion.   In a constant pressure
system, the enthalpy (H) of this system is constant.
It is necessary to balance the combustion heat
released against enthalpy of the combustion products
at their final temperature, which is the
theoretical flame temperature (reactants are assumed
at STP where H is 0).

For example, natural gas (assume all $CH_4$) is
completely burned with 5% excess air.   The balanced
equation for 1 mole of $CH_4$ is:

$$CH_4 + 2.1\ O_2 + 7.9N_2 \longrightarrow$$
$$CO_2 + 2H_2O + 0.1\ O_2 + 7.9N_2 \qquad (8.18)$$

Oxygen remains in the products because excess air was used.

The lower heat of combustion (all products gaseous) of methane at constant pressure is 191.3 K Cal/g mole at 273° (0°C) and 1 atmosphere (from thermodynamic tables of data). The increased enthalpy values for the products at the final temperature can be obtained by (assuming all products to be ideal gases):

$$H = \int_{273}^{T} N_{CO_2}(C_p)_{CO_2}\, dT + \int_{273}^{T} N_{H_2O}(C_p)_{H_2O}\, dT$$

(8.19)

$$+ \int_{273}^{T} N_{O_2}(C_p)_{O_2}\, dT + \int_{273}^{T} N_{N_2}(C_p)_{N_2}\, dT$$

where:  $T$ = theoretical flame temperature, °K
$N$ = number of moles
$C_p$ = specific heat at constant pressure

Values of $C_p$ for these products are:

$$(C_p)_{CO_2} = 18.036 - 4.474 \times 10^{-5}\sqrt{T} - 158.08/\sqrt{T}$$

$$(C_p)_{H_2O} = 6.970 + 0.3464 \times 10^{-2}\, T - 4.833 \times 10^{-7}\, T^2$$

$$(C_p)_{O_2} = 6.732 + 0.1505 \times 10^{-2}\, T - 1.791 \times 10^{-7}\, T^2$$

$$(C_p)_{N_2} = 6.529 + 0.1488 \times 10^{-2}\, T - 2.271 \times 10^{-7}\, T^2$$

$$(C_p)_{CH_4} = 4.750 + 1.200 \times 10^{-2}\, T + 0.3030 \times 10^{-5}\, T^2$$
$$- 2.630 \times 10^{-9}\, T^3$$

(If any $CH_4$ remained unburned, it would be included in the products.) Substitution of the proper values into Equation 8.19 and solving yields the maximum theoretical flame temperature directly.

Values of H can also be obtained for the products at elevated temperatures from thermodynamic tables. The procedure then becomes trial and error as shown here. Estimate that the answer is 2200°K, then:

$$H_{CO_2} = (1)(24,836) = 24,836$$

$$H_{H_2O} = (2)(19,703) = 39,406$$

$$H_{O_2} = (0.1)(15,971) = 1,597$$

$$H_{N_2} = (7.9)(15,158) = \underline{119,800}$$

$$185,639 \text{ cal/g mole}$$

This is not equal to 191,300 (heat of combustion), so guess a higher T of 2300°K and obtain by the same procedure:

$$H = 196,680$$

Interpolate to obtain the answer of T = 2251°K (3600°F), which is the theoretical flame temperature.

## 8.6 REFUSE COMBUSTION

It is estimated that approximately 80% of the 480,000 tons of solid waste generated daily in the U.S. could be disposed of by pyrolysis (*i.e.*, roasting in the absence of air). Treatment of wastes by this manner at 1700°F would produce gas, organic liquid and water vapor, plus some undesirable carbon monoxide, nitrogen oxides and ash. At this time, the process itself does not appear economical as it would cost (after returns from sale of products) about $1.25/ton to carry out the operation. However, incineration plus disposal by land fill costs $8-10/ton, so there could be an actual cash savings if pyrolysis were used in place of incineration.

Another method of refuse combustion reported from a pilot plant of Combustion Power Co. at Menlo Park, California uses a 1500 KW power plant to burn 400 ton/day solid waste. This waste provides about 5-10% of their electrical requirements for 160,000 persons in its burning; a positive step toward conserving our resources, but one that requires proper combustion and air pollution controls to prevent air pollution.

Refuse burning of automobiles in smokeless incinerators, then hand dismantling, can be economically performed at this time. An average auto weighing 3,500 pounds contains approximately 70.1% steel, 14.6% cast iron, 1.0% copper, 1.5% zinc, 1.5% aluminum, 0.6% lead, 4.1% rubber, 2.5%

glass and 4.1% miscellaneous.  It costs $51 to
produce $56 worth of sorted metal products,
including a 19% annual return on money invested
in the equipment.

8.7  THERMAL AND CATALYTIC CONVERSIONS

So far, we have discussed mainly direct or
flame type oxidation.  It is now time to include
thermal (indirect or no flame) and catalytic type
conversions.  These latter two methods of
conversion are primarily used in cases of odor
control and for safety.  Saturated hydrocarbons
are almost odorless.  Usually, the more unsaturated
the hydrocarbon, the more odorous it is.  Alcohols
have low but noticeable odors and aromatics are
highly odorous.  In addition to having odors, many
of these organic compounds are poisonous both to
animals and plants and, for this reason alone,
should not be released into the atmosphere.
Oxidation of these low concentration and poisonous
materials is frequently a desirable method of air
pollution control.

Direct oxidation by flame incineration frequently
is not the most economical procedure.  This is
shown by the cost figures in Table 8.4.  The
disadvantage of direct oxidation is that the
discharge of odorous, poisonous or explosive gases
into burners can result in safety hazards due to
leakage of these gases into the operating room
atmosphere, also explosions can result from
careless mixing.  If the gas to be incinerated is
already at an elevated temperature, this adds a
further advantage to incineration by reducing heat-
up costs and providing added heating value.

TABLE 8.4  OXIDATION CONDITIONS AND COSTS

| Method | Operating Temp., °F | Costs | |
| | | Equipment $/scfm | Annual Fuel, $/1000 scfm |
|---|---|---|---|
| Flame | 2,500+ | 5-10 | 0-20 |
| Thermal (no flame) | 1,000-1,500 | 1.75-10 | 0-7.50 |
| Catalytic | 600-900 | 1.75-5 | 0-4.50 |

Catalysts are useful in speeding up the combustion reaction rate but catalyst activity can be easily impaired. The two mechanisms by which catalysts are poisoned are by chemical reaction with the gas stream and by physical coating. Large industrial catalytic units have been used for many years in the production of various chemicals. For example: thin film platinum wire gauze is used in nitric acid production for the oxidation of ammonia; and vanadium pentoxide ($V_2O_5$) is used for the oxidation of $SO_2$ to $SO_3$ to produce sulfuric acid by the Monsanto Process. Catalysts are substances which change the rate of chemical reaction without themselves being permanently depleted. They may or may not actually enter into the reaction and if they are consumed at all, the rate of consumption is small. Most catalyst loss results from attrition.

Reduction as well as oxidation is required to effectively control air pollution, and *catalytic* reduction is the primary means by which reduction is accomplished. Platinum, for example, can be used to both decolor as well as to abate nitrogen dioxide pollution. Nitrogen dioxide is a reddish-brown gas which colors the atmosphere. If it can be reduced to nitric oxide, which is colorless, it will no longer be visible. $NO_2$ can be decolorized using either hydrogen or methane ($CH_4$) according to the following equations where platinum is the catalyst:

$$NO_2 + H_2 \xrightarrow{Pt} NO + H_2O$$

$$NO_2 + \tfrac{1}{4}CH_4 \xrightarrow{Pt} NO + \tfrac{1}{4}CO_2 + \tfrac{1}{2}H_2O$$

The reaction speeds are fast and all of the products of the reactions are colorless.

Nitric oxide (NO) in the atmosphere can be converted to $NO_2$ by photochemically catalyzed reactions. Therefore, it may be desirable to reduce the $NO_2$ to molecular nitrogen to prevent reformation of the $NO_2$. This is possible by use of platinum (or other) catalysts and at the same time it is possible to reduce NO to elemental nitrogen. The following two types of abatement reactions show how this is accomplished (note that oxygen in the atmosphere also reacts). The first set of abatement equations using hydrogen as the fuel have the limitation of a maximum temperature of 270°F:

$$NO_2 + 2H_2 \xrightarrow{Pt} \tfrac{1}{2}N_2 + 2H_2O$$

$$NO + H_2 \xrightarrow{Pt} \tfrac{1}{2}N_2 + H_2O$$

$$O_2 + 2H_2 \xrightarrow{Pt} 2H_2O$$

The abatement equations to reduce nitrogen oxides using methane fuel have the limitation of a maximum temperature of $234°F$:

$$NO_2 + \tfrac{1}{2}CH_4 \xrightarrow{Pt} \tfrac{1}{2}N_2 + \tfrac{1}{2}CO_2 + H_2O$$

$$NO + \tfrac{1}{4}CH_4 \xrightarrow{Pt} \tfrac{1}{2}N_2 + \tfrac{1}{4}CO_2 + \tfrac{1}{2}H_2O$$

$$O_2 + \tfrac{1}{2}CH_4 \xrightarrow{Pt} \tfrac{1}{2}CO_2 + H_2O$$

The reduction of nitrogen oxides using either hydrogen or methane as fuel is relatively slow while, unfortunately, the reduction of oxygen by both fuels is faster. Although it is possible to obtain concentrations of less than 100 ppm $NO_x$ in the exit stack gas, this process could be expensive if there is a large amount of oxygen to be reduced. New catalysts need to be developed to prevent the reduction of oxygen.

## 8.8   VENTILATION SYSTEMS

It is not intended that industrial hygiene should be covered in this text, but because atmospheric pollutants can be generated in localized atmospheres and then pass into the ambient atmosphere, it is important to at least mention this subject. Ventilation systems are usually designed to exhaust the local pollutants through a stack into the atmosphere. When this occurs, diffusion can be calculated in the standard method using stack gas calculations and atmospheric diffusion equations. If the ventilation system is inadequate to draw the pollutants, they will then diffuse from the building or confined area into the ambient atmosphere through windows or other leakage points. These pollutant concentrations then must be calculated by diffusion equations for line sources on the ground or by the equations which permit calculation of diffusion from a building (see Chapter III).

Certain minimum requirements are needed for adequate industrial ventilation. Figure 8.6 shows a common design of ventilation hood. The canopy hoods must be designed so that the distance from the top of the hood to the vessel is no more than 2½ times the distance from the edge of the vessel to a line even with the edge of the hood. Air movement through the opening into the canopy should be at least 80 ft/min for relatively clean operations. For dusty (*e.g.*, grinding) operations with extremely hazardous chemicals or where large volumes of pollution are released, face velocities as high as 2,000 ft/min are required (7). Velocities this high will even permit some diffusion of molecules into the local atmosphere as is indicated by the kinetic molecular theory of gases:

$$v \simeq \sqrt{\frac{3RT \ g_c}{M}} = \sqrt{\frac{3PV \ g_c}{M}} \qquad (8.20)$$

which gives in English units at STP:

$$v = 8.56 \times 10^3 \ M^{-0.5} \qquad (8.21)$$

The lightest molecule, $H_2$, then has an average velocity of 6,060 ft/sec (363,000 ft/min). The velocity of heavy $SO_2$ molecules is inversely proportional to the square root of the molecular weight ratio but is still 64,300 ft/min.

It is important to note that every cubic foot of air exhausted from a building must be replaced to keep the building from becoming negatively pressurized. If this happens, the exhaust system will not function at the required face velocities. This is true for stacks that draw air from the inside as well as for exhaust and ventilating systems. It is necessary to warm the make up air when it is cold outside. Some typical design criteria are given in Table 8.5 which can also be used to estimate make up air requirements.

Open end ducts can be used to sweep in pollutants from a localized atmosphere. The limitations of these devices are extremely severe, as the inlet velocity drops drastically with increasing distance from the end of the duct. Figure 8.7 shows such an opening and indicates lines of constant velocity at various distances and angles from the face of the opening. The velocity, at various distances perpendicular to

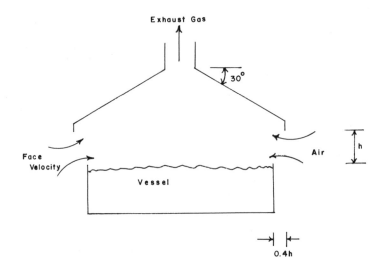

Figure 8.6   Industrial or Laboratory Canopy Type
Hood

the face of the opening, can be found by Equation
8.22 for round, square or rectangular ducts with
length to width dimensions that vary up to 1:3:

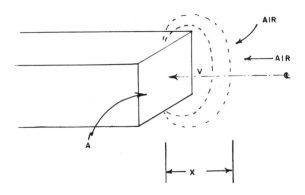

Figure 8.7   Open End Duct Used as a Vent

$$v = \frac{Q}{10\ X^2 + A} \qquad\qquad (8.22)$$

where:   v = velocity along center line, ft/sec
         Q = volumetric flow rate in duct, ft$^3$/ sec
         A = duct face opening area, ft$^2$
         X = perpendicular distance from duct, ft.

The kinetic molecular theory can again be used to indicate the portion of gas molecules that would escape entering the vent because of their kinetic energy. The capture of particulate material will depend on the settling velocity of the particle and the drag by the gases. This is discussed in Chapter IX.

TABLE 8.5   SOME EXHAUST AIR REQUIREMENTS FOR LOW TOXICITY MATERIALS (8)

| System | *Face Velocities, ft/min.* |
|---|---|
| Drying baking curing ovens | 100-200 |
| Melting operations | 200-300 |
| Pickling tank | 150-200 |
| Pulverizer | 130-150 |
| Vacuum filter | 120 |
| Plastic molding machine | 150 |
| Liquid type filter | 100 |
| Screening | 200 |
| Bag filling station | 450-500 |
| Conveyor | 150-200 |
| Hopper ventilation | 150 |
| Process vessel door | 500 |
| Mixer | 200-300 |

QUESTIONS FOR DISCUSSION

1.  Name the pollutants that are released from combustion type operations.
2.  What are the significant differences between coal combustion and fuel oil combustion?
3.  What factors must be considered when evaluating the emissions from combustion type operations?
4.  How would you design chimneys for typical household heating systems using coal, oil, gas and electricity as fuels?
5.  Discuss the relative advantages and the disadvantages, as far as air pollution is concerned, in the use of coal, oil, gas and electricity for residential heating.
6.  How can you tell by appearance whether a power plant is coal or oil or gas fired?
7.  How tall should a stack be and why or why not should stacks be made taller?
8.  If 1,000 foot high stacks were built in Illinois and you lived in central Ohio, what would be your reaction?
9.  What would be the best way to incinerate the refuse in a metropolitan location?
10.  What would be the best way to dispose of refuse in a non-industrial location?

PROBLEMS

8.1  Calculate the volume of stack gas released during the burning of 100 barrels of number 2 grade oil.
8.2  a.  If the oil in Problem 8.1 were burned in one hour, what would be the stack gas exit velocity?
     b.  What should the stack diameter be?
     c.  What should the stack height be?
8.3  a.  If the boiler in Problem 8.1 is a horizontally fired system, what would the stack gas concentration of nitrogen oxides and sulfur dioxide be (use two different methods to obtain each of these answers)?
     b.  Where would the maximum ground level concentration on $SO_2$ and $NO_x$ occur if the effective plume height is 475 feet?
8.4  Rework Problems 8.1 - 8.3 for a coal burning installation that uses five tons of typical Pennsylvania bituminous coal per hour (one method for $NO_x$ and $SO_2$ conc.).
8.5  Frequent daytime concentration of $SO_2$ on the ground one mile downwind from a local coal burning

power plant is 0.015 ppm. If this is a 1,500 hp
facility, what can you conclude about the fuel
consumption rate, plant design (e.g., stack height,
etc.) and methods of operation?

8.6 How much heat is lost per hour because of
excess air when 1000 scfh of natural gas are
burned with 10% excess air (gas temperature ≃
1800°F) and what is the approximate exhaust gas
cmmposition?

8.7 What is the theoretical flame temperature for
the combustion of natural gas (assume 100% $CH_4$)
with 4% excess air if combustion is only 90%
complete (assume no CO formed; all unburned fuel
remains as $CH_4$)?

REFERENCES

1.  Ermenc, E.D., "Controlling Nitric Oxide
    Emissions," Chemical Engineering, Vol. 77, No.
    12, pp. 193-196 (1970).
2.  White, H.J., "Effect of Fly Ash Characteristics
    on Collector Performance," JAPCA, Vol. 5, No.
    5, pp. 37-50 (1955).
3.  Smith, W.S. and C.W. Gruber, "Atmospheric
    Emissions from Coal Combustion--An Inventory
    Guide," U.S. Department of Health, Education,
    and Welfare, Public Health Service Publication,
    no. 999-AP-24 (1966).
4.  Woolrich, P.F., "Methods of Estimating Oxides
    of Nitrogen Emissions for Combustion Processes,"
    American Industrial Hygiene Association Journal,
    Vol. 22, pp. 481-484 (1961).
5.  Smith, W.S., "Atmospheric Emissions from Fuel
    Oil Combustion--An Inventory Guide," U.S.
    Department of Health, Education and Welfare,
    Public Health Service Publication, No. 999-
    AP-2 (1962).
6.  "Industrial Combustion Data," Third Ed., Hauck
    Manufacturing Company--Combustion Engineers,
    168 pp. (1953).
7.  Schuman, M.M., et. al., "Industrial Ventilation
    --A Manual of Recommended Practice," Eleventh
    Ed., Library of Congress, Catalog Card No. 62-
    12929, American Conference of Governmental
    Industrial Hygienists (1970).
8.  Constance, J.D., "Estimating Exhaust-Air
    Requirements for Processes," Chemical
    Engineering, Vol. 77, No. 17, pp. 116-118
    (1970).

# CHAPTER IX

# PARTICULATE COLLECTION THEORY

There are two obvious ways in which air
pollution control methods could be discussed. One
is to present various types of control equipment
and consider how the devices function and the
principles of separation involved. The second
method is to discuss general control theory and then
relate it to specific equipment. It seems more
logical to follow the second procedure because
some devices actually utilize several different
mechanisms for control. Also, with little or no
modification, the same basic principles are utilized
in several different types of apparatus.

This chapter presents the gas cleaning principles
related to particulate collection. Chapter X
discusses gaseous control methods and finally
Chapter XI discusses specific types of control
equipment. Before reading this chapter, it is
suggested that Section 7.1 be thoroughly understood.

Section headings in this chapter do not relate
specifically to equipment type. Instead, the
heading attempts to specify the basic collection
principle discussed. For example, inertial
deposition includes the mechanism by which particles
can be collected by equipment as diverse as cyclone
separators, Venturi scrubbers and filters. Combustion, as a mechanism for removing particulate matter,
is not included in this chapter but is discussed
as a special section in Chapter VIII.

9.1  GRAVITATIONAL SETTLING

A particle in the atmosphere comes under the
influence of three separate forces as shown in
Figure 9.1. The forces are gravity, bouyancy and

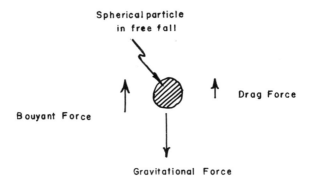

**Fig. 9.1   Forces Acting on a Falling Sphere**

drag.  The force of gravity, which acts downward,
is equal to the mass times the acceleration.  For
a spherical particle, this is:

$$F_G = 4/3 \ \pi r^3 \ \rho_p g \qquad (9.1)$$

where:   $r$ = particle radius
$\rho_p$ = particle density
$g$ = acceleration of gravity

Bouyance, which is attributable to the displaced
air, provides a force which acts upward and is:

$$F_B = 4/3 \ \pi r^3 \ \rho_a g \qquad (9.2)$$

where:   $\rho_a$ = density of air

Drag force acts in a direction opposite the movement
of the particle.  For a falling particle, this is
upward and is obtained using Newton's drag equation:

$$F_D = \frac{C_D \ \rho_a \ (v_p - v_a)^2 \ A}{2} \qquad (9.3)$$

where:   $C_D$ = dimensionless drag coefficient
$A$ = projected area ($\pi r^2$ for a sphere)

A force balance made on the particle at terminal
free fall velocity, assuming the velocity of the
air to be approximately zero, gives:

$$C_D \frac{\rho_a v_p^2}{2} \pi r^2 = 4/3 \ \pi r^3 \ (\rho_p - \rho_a) g \quad (9.1,2,3)$$

Stokes' law for streamlined flow of spheres can be used when the drop Reynolds' number (Re) is low (Re<0.1). For these conditions of viscous Stokes' equation gives:

$$C_D = \frac{24}{Re} \quad (9.4)$$

The drop Reynolds number is equal to:

$$Re = \frac{d \ (v_p - v_a) \ \rho_a}{\mu_a} \quad (9.5)$$

which, at low air velocities, is approximately equal to:

$$Re \simeq \frac{d \ v_p \rho_a}{\mu_a} \quad (9.5a)$$

where:  $d$ = particle diameter
$\mu_a$ = viscosity of air

Substituting for $C_D$ and Re in the force balance, we obtain Stokes' terminal settling velocity equation:

$$v_s = \frac{2r^2 (\rho_p - \rho_a) g}{9 \ \mu_a} = \frac{d^2 (\rho_p - \rho_a) g}{18 \ \mu_a} \quad (9.6)$$

where:  $v_s$ = terminal settling velocity of particle

Remember that this equation was derived using the limitations that the velocity of the air is negligible and the drop Reynolds' number of the falling particle is low. If there is an appreciable air velocity moving in the opposite direction to the particle, it cannot be neglected and must be considered in the calculations. This terminal settling velocity equation provides reasonably approximate values for Reynolds' numbers up to 2.0. (If any gas other than air is used, values appropriate to that gas must be used.)

At Reynolds' numbers greater than 2, the value of the drag coefficient must be determined empirically. For intermediate Reynolds' numbers of $2 < Re < 10^3$, the drag coefficient can be calculated according to:

$$C_D = (18.5)(Re)^{-0.6} \qquad (9.4a)$$

At significant increases in Reynolds' numbers, it should be noted that the particle will no longer be in free fall. This will occur somewhere in the intermediate Reynolds' number region.

For turbulent regions which occur at high Reynolds' numbers of $10^3 < Re < 2 \times 10^5$, the drag coefficient, according to Newton's law, can be assumed to be constant:

$$C_D = 0.44 \qquad (9.4b)$$

When the particles have a very small diameter, they tend to "slip" between the air molecules and, as a result of this, fall faster. When this happens, Stokes' equation must be modified by a correction factor as follows:

$$C_D = \frac{24}{C\ Re} \qquad (9.4c)$$

where:  $C$ = Cunningham correction factor, dimensionless

The corrected terminal settling velocity then becomes:

$$v_s(\text{corrected}) = v_s\ C \qquad (9.6a)$$

The Cunningham correction factor can be approximated using the following empirical equation:

$$C = 1 + \frac{3.45 \times 10^{-4}T}{d} \qquad (9.7)$$

where:  T = absolute temperature, °R
       d = particle diameter, μ

This correction can be significant for particles $\leqslant 1\mu$ in diameter.

Figure 3.27 is a curve of terminal settling velocity versus particle diameter for spherical particles with a specific gravity of 2.0. This curve is corrected for the slip of the small diameter

particles and the drag variation of the larger
diameter particles because of the changes in Re.
    There are times when the particle cannot be
considered as spherical. When this occurs, it is
necessary to account for the deviation from
sphericity. Section 7.1.5 discusses some ways for
accounting for this. If the sphericity factor (K')
calculated by the method of 7.1.5 is used, then the
Stokes' equation for terminal settling velocity in
regions of viscous flow becomes:

$$v_{(s)} = K' \frac{d^2(\rho_p - \rho_a)g}{18\,\mu_a} \qquad (9.6b)$$

for viscous flow where Re < 0.05.
    Mechanisms which utilize variation of gravita-
tional settling are elutration and sedimentation.
Air elutration is utilized to separate particulate
samples into their various size fractions. This is
done by varying the velocity of the upward air
current and can be used to separate particles with
diameters as large as about 40 microns. Equation
9.6 corrected for drag coefficient variation and
*including* upward *air velocity* can be used to
calculate the maximum diameter of particles removed
at various given air velocities.
    Sedimentation is usually carried out in a liquid
fluid (instead of air) because settling rates of
particles in the liquid are much lower making the
separation easier. The settling rates are valid
for Reynolds' numbers less than 0.2. Equation
9.6 can again be used when corrected as necessary.
Quite frequently, it is possible to neglect the
density of air in Equation 9.6. It is not possible
to neglect the liquid density during sedimentation
work. The minimum diameter for effective grading
by sedimentation work is approximately 3 microns.
The limiting maximum diameter to keep within the
specified Reynolds' number varies depending on the
density of the particulate. For example, limits in
water are:

|  | $\rho$, *g/cc* | *Limiting Diameter*, $\mu$ |
|---|---|---|
| Silica | 2.5 | 50 |
| Hematite iron ore | 5.0 | 36 |
| Lead | 11.4 | 26 |

## 9.2  INERTIAL DEPOSITION

### 9.2.1  IMPACTION

A particle in motion tends to remain in motion just as a particle at rest tends to remain at rest. In other words, they resist changes. Inertia is the change of motion and is equal to the force which is attempting to create a change times the time through which it acts, or:

$$\text{Inertia} = \int d(mv) \qquad (9.8)$$

Inertial impaction then becomes the collection of moving particles by impinging them on some "target" which may be stationary or moving at a different velocity. Particles in a moving gas stream travel at approximately the same velocity as the gas. Because it has a low mass, the gas moves around the target as shown by the gas streamlines of Figure 9.2. Particles, however, with their high density, have more inertia and therefore, resist changes in direction. As a result, they can travel as shown in Figure 9.2, in a straighter line and may hit the target. The larger the particle the greater the inertia and the less the particle will tend to change direction. This means that some small particles, at a given velocity, will move around the target while large particles will strike the target. Particles that hit the target do not necessarily stay there. They may bounce off and return to the air stream or they may be knocked off the target by other particles which strike them.

An approximate generalized target efficiency equation for a simple one-stage device, such as a filter bed, cyclone, etc. is:

$$\eta = \exp-\left[\frac{0.018}{R}\,\psi^{0.5+R} -0.6R^2\right] \qquad (9.9)$$

where:  $\eta$ = effective target efficiency, fraction
$R = d/D_c$
$D_c$ = collector diameter
$\psi$ = dimensionless impaction parameter

$$=\left(\frac{\rho_p\,v_o\,d^2}{18\,\mu D_c}\right)^{1/2}$$

$v_o$ = velocity of particle relative to target

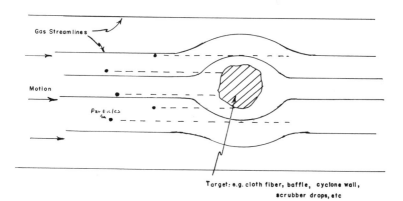

Fig. 9.2    Inertial Impaction--Collecting Particulate
Matter on a Target

This equation is good for drop Reynolds' numbers in
the range of $0.04 < Re < 1.4$; $\eta$ is a constant equal
to 0.15 at values of $\Psi$ below 0.15. The collector
diameter may be anything from the diameter of a
filter fiber to the diameter of a cyclone separator,
because both centrifugal deposition as well as fiber
collection are forms of inertial impaction.

### 9.2.2  STOPPING DISTANCE

Certain devices operate by eliminating the
inertia of the particulate matter. Particles
entrained in a high velocity gas stream can be
introduced into an expansion zone. Upon expansion,
the gas velocity decreases to approximately zero.
The particle is then under the influence of a drag
force which tends to slow it down. The distance that
it takes to stop the particle is a function of the
initial inertial energy of the particle and the
drag resistance. (It is necessary to consider this
stopping distance when designing inertial impaction
collectors to be sure that the particle does not
stop before striking the target.) Figure 9.3 shows
gas and particulate matter entering at the left of
the diagram at a velocity of $v_g$. Upon entering
the expansion zone, the gas velocity becomes
approximately equal to zero while the particles have
an initial velocity $v_i$ (which is approximately equal to
$v_g$). The particle is slowed down by the drag force
and the distance that it takes to stop the particle

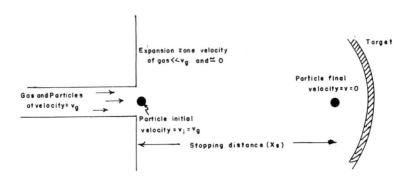

Fig. 9.3  Particle Stopping Distance

is shown as $X_s$. Neglecting the force of gravity temporarily, the remaining forces, which consist of Newton's drag force, can be equated using an average velocity $(v_p)_{av}$ to give:

$$-\frac{dv_p}{dt} = \frac{9\ \mu_a (v_p)_{av}}{2\ \rho_p\ r^2}\qquad(9.10)$$

Integrating from $v_i$ to $v = 0$ and realizing that the stopping distance equals the average particle velocity times the time, the Stokes' stopping equation is obtained:

$$X_s = \frac{2\ v_i\ \rho_p r^2}{9\ \mu_a}\qquad(9.11)$$

This equation is valid for spherical particles in horizontal motion being slowed down by drag force only; forces such as gravitational, magnetic, electric and thermal forces are not included.

   The slip of the particles through the air which occurs in the same manner as discussed in Section 9.1, may make it necessary to use the Cunningham correction factor.  The corrected stopping distance then becomes:

$$X_s(\text{corrected}) = X_s\ C\qquad(9.11a)$$

Figure 9.4 is a plot of Stokes' stopping distance versus initial velocity in air for spherical

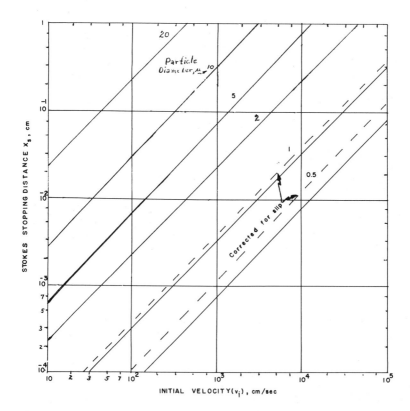

Fig. 9.4   Stokes Stopping Distance versus Initial
           Velocity in Air at SC  For 1 g/cm³
           Particles

particles moving perpendicular to gravity.  This is
for particles with a density of 1 gram/cc.  The
curves for particles less than 2 microns in diameter
include the Cunningham correction factor to account
for slip.  It can be seen that the 0.5 micron
diameter particles show a considerable difference
in stopping distance when corrected.  Stopping
distances for particles other than 1.0 gram/cc
density can be obtained by multiplying the values
from Figure 9.4 by the density of the material
being considered.

It is necessary to account for the force of gravity before being able to establish the final position of the particle. If the particle was initially moving perpendicular to the gravitational force, the final drop that the particle would experience in the stopping time must be vectorially added to obtain the final position. If the particle is moving parallel with the gravitational force, this influence must be added to or subtracted from the initial equations in order to derive an equation that will reflect a proper stopping distance.

## 9.2.3   CENTRIFUGAL DEPOSITION

Centrifugal force is one of the most commonly used methods of affecting a separation of particulate matter from gas streams. In a cyclone separator, both gravity and centrifugal force are present. Equation 9.6, as corrected for slip if necessary, shows terminal settling velocity ($v_s$). The terminal centrifugal velocity in the normal direction ($v_n$) can be calculated by a similar equation:

$$v_n = \frac{d^2 \, \rho_p \, v_t^2}{18 \, \mu_a r_s} \qquad (9.12)$$

where:   $v_t$ = tangential velocity of gas
$r_s$ = radius of streamline

The tangential gas velocity is approximately equal to the inlet gas velocity and the streamline radius depends on where the particle is with relation to the center of the cyclone. The separation factor for a cyclone separator is an indication of how the centrifugal force compares to the force of gravity. Separation factor is equal to the ratio of the terminal free fall velocity ($v_s$) and can be written:

$$\text{separation factor} = \frac{v_t^2}{g \, r_s} \qquad (9.13)$$

For example, a cyclone operating with a 4000 ft/min inlet gas velocity with a particular streamline radius of 4 inches has a separation factor of 416. A spherical ten micron particle with a specific gravity of 2.7 falls at the rate of 1.6 ft/min but the rate of the particle traveling toward the wall due to the centrifugal velocity is 666 ft/min.

The separation factor can be increased by increasing the terminal velocity and decreasing the radius of the streamline.

## 9.3  DIFFUSION OF PARTICLES

The first figure in this book, Figure 1.1, shows a schematic representation of generalized collection efficiency versus particle diameter.  This curve indicates that there is a distinct size difference between those particles effectively captured by inertial collection and those captured by diffusion. Figure 1.1 indicates that the most difficult size particles to collect are those between the two regions (approximately 0.2-1.0 μ diameter).  As particle diameter decreases, the surface to mass ratio increases.  When the particles become sufficiently small, they can be bombarded by gas molecules and forced to move like molecules in random directions. This is known as Brownian motion.

The rate at which a particle diffuses can be calculated by equations such as the Stokes'-Einstein equation:

$$D_{PM} = \frac{CKT}{3 \pi \mu d} \qquad (9.14)$$

where:  $D_{PM}$ = diffusivity of particle through continuous medium, $cm^2/sec$
$C$ = Cunningham slip correction factor (Eq. 9.7), dimensionless
$K$ = Boltzman constant = $1.38 \times 10^{-16}$ $\frac{g \ cm^2}{sec^2 molecule \ °K}$
$T$ = absolute temperature, °K
$\mu$ = medium viscosity = $1.8 \times 10^{-4}$ g/(cm sec) for air at SC

This equation applies for both small spherical particles as well as large molecules.  The Stokes'-Einstein equation can also be used for particles or molecules in liquids.  Figure 1.1 shows that the efficiency increases in the small diameter region as diameter decreases.  This suggests that increased particle diffusivity accounts for this collection efficiency increase.  An example of diffusivity in air for several small diameter particles and for $SO_2$ molecules is given in the following list at SC (gaseous $SO_2$ molecules have a diameter in the range of $3 \times 10^{-4}$ microns):

| *Particle* *d, micron* | $D_{PM}$, $cm^2/sec$ | *Schmidt, No.,* $\mu/\rho D_{PM}$ |
|---|---|---|
| 0.5 | $6.4 \times 10^{-7}$ | $2.4 \times 10^5$ |
| 0.1 | $6.57^{-6}$ | $2.30 \times 10^4$ |
| 0.01 | $4.44 \times 10^{-4}$ | $2.41 \times 10^2$ |
| 0.001 | $4.22 \times 10^{-2}$ | 3.59 |
| $SO_2$ molecules | $1.18 \times 10^{-1}$ | 1.28 |

Collection of small diameter particles by diffusion is, in a sense, an impaction mechanism because the particles are captured on a target. In addition to their diffusional movement, particles traveling in a moving gas stream have a net direction velocity. Diffusion occurs in all directions, yet it is possible to obtain a pseudo-diffusional veolcity by dividing the diffusivity by the volume and multiplying by the cross section area normal to the chosen direction of flow. Although directional velocity due to the moving gas stream $(v_p)$ increases inertial collection, it causes a reverse effect and decreases diffusional collection. Direction velocity is usually the same as the gas stream velocity. The relative velocity $(v_o)$ is the difference in velocity between $v_p$ and the target which may or may not be stationary. Johnstone and Roberts [1] provide equations for predicting diffusional collection efficiency for cylindrical and sperical type targets using relative velocity. The equations are for Schmidt numbers $(N_{Sc})$ less than $2.4 \times 10^5$ (particles less than $0.5\mu$ in air):

For cylinders: $\eta = \dfrac{D_{PM}}{D_c v_o} \left[ \dfrac{1}{\pi} + 0.55 \ N_{Sc}^{1/3} \ N_{Re}^{1/2} \right]$

$$(9.15)$$

For spheres: $\eta = \dfrac{D_{PM}}{D_c v_o} \ 2\left[ + 0.55 \ N_{Sc}^{1/3} \ N_{Re}^{1/2} \right]$

$$(9.16)$$

where:  $\eta$ = effective diffusional efficiency, fraction
$D_{PM}$ = diffusivity (Equ. 9.14), cm /sec
$D_c$ = diameter of collector, cm
$v_o$ = relative velocity, cm/sec
$N_{Sc}$ = Schmidt No. = $\dfrac{\mu}{\rho \ D_{PM}}$ , dimensionless

$\mu$ = medium viscosity, g/(cm sec)
$\rho$ = medium density, g/cm$^3$
$N_{Re}$ = Reynolds' No., dimensionless

Gases in the atmosphere can also nucleate. For example, $SO_2$ forms groups of 3.5 (average) or more molecules which have diameters of 0.001 microns and larger. $SO_2$ half-life in the atmosphere is usually 8 to 10 hours but may be as long as a year.

## 9.4 AGGLOMERATION

It is possible for particles to become converted from small diameter type which are collected by diffusion to larger size particles which can be collected by impaction. Aerosols do not remain the same size as when they are emitted but "grow" by agglomeration. When small particles move by diffusional or mechanical turbulence, they strike each other and, upon striking, sometimes remain connected, decreasing the total number of particles in a given sample. (It should be noted that solids agglomerate and liquids coalesce.)

The rate of agglomeration depends on concentration and is approximately independent of size. The kinetics of agglomeration are similar to those of a second order chemical rate equation.

$$- \frac{dN}{dt} = kN^2 \qquad (9.17)$$

where:  N = number of particles per unit volume at time t
k = rate constant

The integrated form of this equation, when integrated from time zero when the concentration of particles is $N_o$ to time t, is:

$$\frac{1}{N} - \frac{1}{N_o} = kt \qquad (9.17a)$$

Rate constants for agglomeration can be estimated for both homogeneous and heterogeneous systems in the following manner:

For homogeneous:  $k = \frac{4 K T C}{3 \mu}$ (9.18)

For heterogeneous:  $k \simeq 1.5$ times Equation 9.18
where:  $k$ = rate constant, cm$^3$/sec

The value of the rate constant for homogeneous
systems in air at 20°C is $3 \times 10^{-10}$ cm$^2$/sec.

Large particles increase in size at the expense
of the smaller particles. for this reason, fog
droplets are usually found to be greater than five
microns in diameter. Table 9.1 shows that particles
with a concentration greater than $10^7$ particles/ft$^3$
agglomerate in a matter of a few minutes. Therefore,
concentrations greater than this are unstable in
the atmosphere and are usually not found because it
usually takes longer to measure the size distribution
than it does for agglomeration to create a new size
distribution.

TABLE 9.1   PARTICLE AGGLOMERATION RATES

| Number of Particles Per Ft of Air at S.C. | | Agglomeration Time to Form Final Concentration from Initial |
|---|---|---|
| Initial Conc. | Final Conc. (1/10 initial) | |
| increas- $10^{10}$ | $10^9$ | 3 sec |
| ingly $10^9$ | $10^8$ | 0.5 min (30 sec) |
| unstable $10^8$ | $10^7$ | 5.0 min |
| rela- $10^7$ | $10^6$ | 50.0 min |
| tively $10^6$ | $10^5$ | ~9 hr (500 min) |
| stable | | |

Values of the rate constants can also be
increased by increasing the pressure and by
increasing turbulence. Increased pressure reduces
the distance between the molecules making it easier
for increased collision and the resulting
agglomeration. Greater turbulence increases the
frequency at which the particles strike each other.
Turbulence can be increased by both mechanical
methods and by the introduction of energy, as for
example, in the form of sonic waves. The rea-tion
rate will be increased by raising the temperature
and lowering the viscosity as is obvious from
Equation 9.18.

The diameter of particulate matter can change
by factors other than agglomeration and coalescence.
Hygroscopic liquid particles (for example, sulfuric
acid) are smaller and more concentrated at low
humidity than at high humidity conditions. As the
humidity is increased, the number concentration of
particles decreases and the diameter increases
because of deliquescence.

## 9.5  ELECTROSTATIC ATTRACTION

The fact that unlike charges attract each other provides a mechanism that forces particles to migrate out of a gas stream and become deposited on the surface of a collector. This precipitation by electrostatic forces depends on four factors:

1. The pollution in the gas stream must contain particulate matter.

2. A discharge electrode (which is usually a negatively charged wire) must be located in the center of the moving gas stream.

3. The collecting electrodes must have large surface area--they are usually plates or duct walls and at ground potential (which means they are positive with respect to the discharge electrode).

4. A high potential DC electric field of approximately 70,000 volts is necessary.

At a critical voltage, air molecules become ionized at the discharge electrode and dissociate into positive and negative ions. This corona of gaseous ions is evident at the discharge electrode by a low intensity glow which can be seen in a darkened room. No sparks should occur under normal operation. The negative air ions are attracted toward the collector. As they move, they attach themselves to the pollution particles giving the particles a negative electrical charge which causes them to be attracted to the collectors. Particle charging occurs very fast and, as a result, most of the charging is done near the entrance of the precipitator. The amount of charge that a particle can hold is proportional to the particle diameter squared. For example, a 1 micron particle can obtain a 200 esu charge and a 10 micron particle can hold a 20,000 esu charge. (Esu means electrostatic unit, which is $1.6 \times 10^{-19}$ coulombs; 1 coulomb = $6.2 \times 10^{18}$ electrons.) Figure 9.5 shows schematically how particles of dirt are charged and then migrate to the collecting electrodes.

The negatively charged ions have a relatively long distance to travel before they reach the collector while the positive ions are kept concentrated near the discharge electrode. This makes it possible to contact, charge (negatively) and deposit an overwhelmingly large number of the total particles. Some of the particles that are near the discharge wires do become positively charged and are attracted and move to the wires. This occurs for only a small percentage of the particles. The

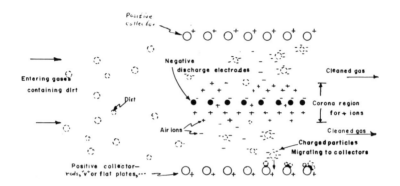

Fig. 9.5  Electrostatic Charging

fractional collection efficiency for electrostatic
attraction and precipitation ($\eta$) may be as high as
0.999+.

Charged particles become neutralized when they
reach either the collector or the discharge wire.
Some of the particles that reach the discharge
electrode become re-entrained in the stream, but
those that stick to the wire must be mechanically
removed from the unit periodically.  The particles
that reach the collector can also become re-entrained
but this is undesirable.  The material that sticks
to the collector is removed by mechanical methods
such as rapping, washing, etc.

Electrostatic collection has the advantages of
being extremely efficient as well as being able to
handle high temperature gases.  A disadvantage is
that many types of particulate material are not
suited for electrostatic collection because:  the
material will not hold a charge; the stickiness
of the material results in subsequent clogging
of the device; and/or the material has a high bulk
resistivity.  The greatest use of electrostatic
precipitators is in the recovery of fly ash.  The
industrial use of electrostatic precipitators can
be broken down approximately according to the
following recovery uses:  60% power plants (fly
ash recovery), 16% metals and metal oxides
recovery, 10% cement dust recovery, 6% paper
industry collection, 3% chemical industry collection
and 5% for other industrial uses.

9.5.1 PARTICLE CHARGING--FIELD STRENGTH AND
VOLTAGE POTENTIAL

Factors which contribute to the charging of
particles in electrostatic precipitators include
variables such as precipitator configuration, voltage
potential, type of gas atmosphere (including humidity
and dust-loading factors), particle size and type of
particulate matter. An easy way to visualize the
mechanism of charging a particle in space is to use
cylindrical coordinates. A cylindrical type of
electrostatic collector is shown in Figure 9.6.
The downstream direction of gas flow is Z and
the distance from the center of the precipitator is
represented by r. $R_1$ is the outside radius of the
discharge electrode and $R_2$ is the inside radius of
the collector tube. The potential difference or
voltage is represented by V and is a scaler quantity.

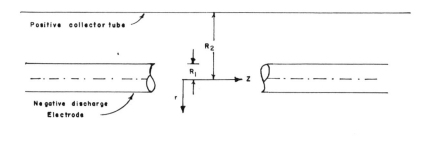

Fig. 9.6  Particle Charging in a Cylindrical Collector

The Gradient of V results in the negative of the
vector quantity E, which is field strength. The
Gradient or Grad V is written as ∇V in vector notation and is similar in significance to the familiar
temperature gradient. (The reader concerned about
vector notation and vector analysis may refer to
such texts as Ref. 2.) Grad V in cylindrical
coordinates is:

$$\nabla V = \frac{\partial V}{\partial r} + \frac{1}{r}\frac{\partial V}{\partial \Theta} + \frac{\partial V}{\partial Z} = \frac{dV}{dr} = -E \quad (9.19)$$

Θ is the angle that can be swept from 0 to 360°
perpendicular to the direction of gas flow. The

voltage change with respect to $\Theta$ is zero because this is a symmetrical system and the change of voltage with a change in distance in the Z direction is 0 for this continuous system. For these results, $\nabla V$ is equal to the total derivative of the change of voltage with respect ot distance in the radial (r) direction.
The voltage difference can then be written as:

$$V = - \int_{R_1}^{R_2} E \, dr \qquad (9.20)$$

Poisson's vector equation for the divergence or dot product of the field strength (which is written $\nabla \cdot E$) is:

$$\nabla \cdot E = \frac{\omega}{K_o} = - \frac{E}{r} \qquad (9.21)$$

where:   $\omega$ = space charge (amt. of charge in a given volume of space
$K_o$ = dielectric of a vacuum
$= 8.8 \times 10^{-12}$ coulombs$^2$/(joule m)

For this cylindrical system, $\nabla \cdot E$ is also equal to:

$$\nabla \cdot E = \frac{\partial E_r}{\partial r} + \frac{1}{r} \frac{\partial E}{\partial \Theta} + \frac{\partial E_Z}{\partial Z} = \frac{dE}{dr} \qquad (9.22)$$

The terms for the change in field strength with respect to $\Theta$ and the change in field strength with change in downstream direction both equal 0 in Equation 9.22 (just as in 9.21) because of symmetry and continuity respectively.
The right hand sides of Equations 9.21 and 9.22 can be equated to show that the ratio of field strength to radial distance is equal to the negative of the first differential:

$$\frac{E}{R} = - \frac{dE}{dr} \qquad (9.23)$$

Rearranging this equation and integrating without limits gives an expression for field strength:

$$\int \frac{dE}{E} = - \int \frac{dr}{r} \qquad (9.24)$$

$$\ln E = -\ln r + \ln C_1 \qquad (9.25)$$

$$E = \frac{C_1}{r} \qquad (9.26)$$

where:   $C_1$ is the constant of integration

If we substitute the value of E from Equation 9.26 into Equation 9.20 and integrate the resulting expression between the limits of $R_1$ and $R_2$, we obtain an expression which enables us to calculate the value of the integration constant:

$$V = \int_{R_1}^{R_2} \frac{C_1}{r} \, dr = C_1 \ln \frac{R_1}{R_2} \qquad (9.27)$$

or:
$$C_1 = \frac{V}{\ln R_1/R_2} \qquad (9.28)$$

Placing this value for the integration constant into Equation 9.26 gives an expression for E in terms of voltage difference, radial direction and the physical dimensions of the discharge electrode and collector tube:

$$E = \frac{V}{r} = \ln \frac{R_2}{R_1} \qquad (9.29)$$

This theoretical equation shows that E increases directly with V and decreases with r.

Equation 9.29 can be rearranged to show voltage potential in terms of field strength which makes it possible to calculate the voltage required to produce the desired ionization without creating arcing. Arcing, which occurs at high voltage, produces ozone gas and is a wasteful expenditure of electrical energy. Field strength values can also be calculated by the empirical equation:

$$E_s = 3.1 \times 10^6 \; b \; 1 + \frac{0.0308}{\sqrt{R_1'}} \qquad (9.30)$$

where:   $E_s$ = field strength at the wire surface,
              volts/meter
         b = smoothness factor (increases from
              0.93 for oxidized or weathered wires
              to 1.0 for smooth polished wires.)
         $R_1'$ = discharge electrode radius, meters

This equation is good when the gas is air at 25° and one atmosphere pressure. Values of the smoothness factor can be obtained for cylindrical wire and tube type systems from Peek (3). In large commercial units, the discharge electrodes are about 1/16 inch diameter, so $R_1 \simeq 1 \times 10^{-3}$ m.

Rearranging Equation 9.29 gives the voltage required at the surface for good corona ionization and particle charging as:

$$V_s = E_s R_1' \ln R_1/R_2 \qquad (9.31)$$

It should be noted that if the collector is too close to the discharge wire, arcing will occur before corona ionization. To prevent this it is necessary to either decrease the discharge wire diameter or increase the spacing distance. A field strength of approximately 3,000 KV/meter is required to produce a corona discharge in dry air at SC from a large surface. As surface area decreases, $E_s$ increases and $V_s$ decreases. Hot gases ionize easier which limits electrostatic precipitator gas temperature to a maximum of approximately 1,500°F for air.

## 9.5.2   FIELD STRENGTH AND CURRENT

The field strength can also be calculated by measuring the amount of current leaving the discharge electrode and by knowing the geometrical configuration of the precipitator. For large tubular precipitators, (where $R_2$ is greater than 1 inch) and for values of field strength of approximately 1000 volts per centimeter, Gottschlich (4) presents a simplified version of the Townsend equation:

$$E = \left[\frac{i}{2\pi K_o K}\right]^{1/2} \qquad (9.32)$$

where:   i = current leaving the discharge electrode, amp/meter of lineal length
$K_o$ = dielectric of vacuum
$K$ = ion mobility, $m^2/(\text{volt sec})$, (about $2 \times 10^{-4}$ for air)

When wire and plate precipitators are used, this equation can be rewritten as:

$$E = \left[\frac{2i\,S}{\pi K_o Kh}\right]^{1/2} \qquad (9.33)$$

where:   S = distance between wires
h = distance from wire to plate

Appropriate values of field strength obtained by
Equation 9.32 or 9.33 can be substituted into
Equation 9.29 to calculate the required voltage
potential. This voltage must be greater than the
voltage required to start a corona as calculated
from Equation 9.31.

Increased concentration of particulate matter
(dust loading of the gas stream) causes a reduction
in the current flow because of the increased value
of the space charge ($\omega$). Equation 9.21 shows that
when space charge increases, this produces a corresponding decrease in field strength at any given
location. Equations 9.32 and 9.33 show that the
current flow is porportional to the square of the
field strength; therefore, as field strength
decreases, the current flow is also reduced.

## 9.5.3 ELECTROSTATIC FORCE

Most of the particles removed in an electrostatic
precipitator are removed as a result of having been
charged by the bombarding negative ions moving
toward the collector. (Particles less than 2 microns
in diameter, however, are charged by ion diffusion,
the rate of which can be estimated by the equations
in Section 9.3.) The amount of charge that accumu-
lates on a particle in an electrostatic precipitator
is proportional to the dielectric of the system,
the field strength, the surface area of the particle,
the ion mobility and the time available for
accumulating the charge.

The dielectric of a system is the resistance of
the system to being broken down and conducting
electrons by current flow. The dielectric ($D_L$) has
the units in the cgs system of coulombs$^2$ joule$^{-1}$
m$^{-1}$ and can be found by:

$$D_L = \varepsilon K_o \qquad (9.34)$$

The dielectric constant ($\varepsilon$) is defined by the
equation:

$$\varepsilon = \frac{q_1\ q_2}{F\ S^2} \qquad (9.35)$$

where:  $\varepsilon$ = dielectric constant of medium
        $q$ = charges on a surface
        $F$ = force of attraction
        $S$ = distance between charged surfaces

Table 9.2 presents dielectric values for a vacuum, air and two typical pollutants.

TABLE 9.2  DIELECTRIC CONSTANTS AND DIELECTRIC OF SEVERAL MEDIA

| *Material* | $\varepsilon$ *at S.C.* | $D_L$, *Coulombs²/ (joule m)* |
|---|---|---|
| vacuum | 1 | $8.8 \times 10^{-12}$ (also $= K_O$) |
| dry clean air | 1.0054 | $8.8 \times 10^{-12}$ |
| calcium carbonate | 6.14 at 10 Hz | $5.4 \times 10^{-11}$ |
| ammonia (liquid) | 17 | $1.5 \times 10^{-10}$ |

9.5.4  CHARGING EFFICIENCY, ELECTRIC WIND AND GAS VELOCITY

The electrostatic charging efficiency (which is not the electrostatic precipitator collection efficiency) varies depending on the turbulence of the gas in the system. This turbulence is created by the geometrical configuration of the precipitator, high flow rates and "electric" (ionic) wind. The electric wind is caused by motion of both the dust and ions moving toward the collector. The electric wind has a terminal drift velocity toward the collector which can be estimated theoretically by the following equation which is derived assuming spherical particles; Reynolds' number resulting from drift is <0.1; particles are fully charged by ion bombardment; and the precipitator has a single charging and precipitating zone so the field strength is the same for both:

$$v_w = \frac{pd'(E_{R_1})^2 c}{3.77 \times 10^{12}\ \mu} \qquad (9.36)$$

where: $v_w$ = electric wind drift velocity, m/sec
$\quad\quad\quad d'$ = particle diameter, meters
$\quad\quad\quad E_{R_1}$ = field strength at discharge electrode, volts/meter
$\quad\quad\quad c$ = Cunningham correction factor, dimensionless
$\quad\quad\quad \mu$ = gas viscosity, g/(cm sec)

$$p = \frac{3\varepsilon}{\varepsilon+2} \cong 1.75 \text{ for low dielectric}$$

materials

$\varepsilon$ = dielectric constant of aerosol particles

Gottschlich predicts that collector electro-static fractional efficiency ($n_e$) is:

$$n_e = 1 - \exp\left[-\frac{A_s v_w}{Q}\right] \qquad (9.37)$$

where:   $A_s$ = collector surface area, $m^2$

   $Q$  =  volumetric flow rate, $m^3$/sec

Figure 9.7 shows electrostatic collection efficiency as a function of gas velocity and electric wind. (Note that the units of the parameter, electric wind, are in English units of feet per second.) This figure shows that collection efficiency increases with increase in electric wind velocity and decreases with increase in gas stream velocity. There is also a critical gas stream velocity which, if exceeded, will cause collected particles to be re-entrained into the gas stream, resulting in a drastic reduction in collection efficiency. Particles larger than approximately 74 microns in diameter are not affected by a critical velocity, however, most of the material collected is smaller than 74 microns in diameter. Examples of several critical velocities are given in Table 9.3.

TABLE 9.3   TYPICAL ELECTROSTATIC COLLECTOR CRITICAL VELOCITIES

| *Material* | *Critical Velocity, ft/sec.* |
|---|---|
| Carbon black | 2 |
| Fly ash | ~8 |
| Cement dust | 10-12 |
| Liquids | 20 |

The magnitude of the electrostatic forces acting under "typical" precipitator conditions is shown in Figure 9.8. This electric force ($F_e$) shown as gravitational units for various diameter particles is calculated by:

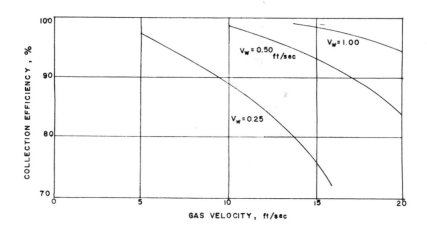

Figure 9.7   Electrostatic Collection Efficiency as
Function of Gas Flow Velocity and
Electric Wind

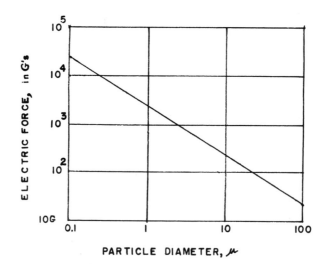

Figure 9.8   Typical Electrostatic Force

$$F_e = q\ E \qquad (9.38)$$

where: q = accumulated charge on particle
(about 25,000 electrons max. on 10
micron particles).

Figure 9.8 can also be used to estimate the
accumulated charge on a given diameter particle.
A summary of the variables which affect
electrostatic collection efficiency shows that the
efficiency can be increased by: 1) decreased
particle dielectric, 2) decreased gas velocity,
3) decreased gas viscosity, 4) increased electric
wind, 5) increased collector area, 7) increased
field strength (or voltage difference) up to where
arcing begins, 7) greater radii of curvature on
particles (smaller particles) and 8) increased
dust loading up to a saturation level (collector
must be cleaned frequently to prevent re-entraining
and blocking).

## 9.5.5 PARTICLE RESISTIVITY

Electrical resistivity is the reciprocal of
conductivity (conductivity is the quantity of
electricity transferred across a unit area per unit
potential gradient per unit of time). Bulk
resistivity ($R_b$) has the units of ohm-centimeter
and is found by:

$$R_b = \frac{AV}{XI} \qquad (9.39)$$

where: A = cross sectional area of dust layer
measured perpendicular to X, $cm^2$
I = current, amp
V = voltage difference
X = dust thickness, cm

The resistivity of industrial precipitates
ranges from $10^{-3}$ ohm centimeter for carbon black
to $10^{14}$ ohm centimeter for dry lime rock dust at
200°F. Dust that is removed in an electrostatic
precipitator builds up on the collector surfaces.
This results in a resistance to the flow of electrons
that are released when each new particle deposited
on the dust in the collector attempts to become
grounded.
Particulate matter is divided into three
resistivity classes. Materials in the low resistivity class have values ranging from the $10^{-3}$ to

$10^4$ ohm centimeter. This means these particles have
a high conductivity that enables a rapid passage of
electrons and a subsequent rapid neutralization of
collected material. When this happens, there is
no electrical force of attraction to hold the
material to the collector surface. When molecular
attraction is weak, as it frequently is, the
material escapes back into the gas stream where it
is recharged, recollected and redeposited numerous
times. Material that does this frequently escapes
collection and passes through the precipitator.
The coarse fraction of boiler flue dust is an
example of this type of material--it is hard to
collect and, because of this, cyclone precleaners
are required before the electrostatic precipitator.

Particulate matter that has a high resistivity
($2x10^{10}$ to $10^{14}$ ohm centimeter) has low conductivity
to the flow of electrons. This results in a large
voltage drop across the dust on the collector
surface. Figure 9.9 shows voltage drop as a function
of distance from the wire for this type of material
with the discharge wire and the collecting plate
pictorially represented on the diagram. Voltage drop
across the collected dust in the example shown
amounts to 20% of the total voltage drop. This
means that only 80% of the voltage is available
to drive the charged particles to the collecting
plate. The voltage drop on the plate varies with
the thickness of the dust as well as the resistivity
of the dust. Collected material having a high
resistivity causes what is known as "back corona"
or "back ionization". Back ionization occurs when
pockets of air trapped in the dust become ionized
because of the high voltage potential across the
dust layer. (If the resistivity were not high,
there would not be sufficient voltage drop to cause
ionization of the gas.) The ionization forms
positive and negative ions. The positive ions
migrate *away* from the collector neutralizing charged
particles and even charging particles positively.
This effectively stops the flow of current from the
discharge wire (stops particle deposition) even
though the voltages are increased as high as
possible. Under this condition, the precipitator
fails to collect any more dust.

The remaining class of particles has an
intermediate resistivity of from $10^4$ to $2x10^{10}$ ohm
centimeter. Material with this range of resistivity
is most suitable for electrostatic precipitation.
It is possible, by proper treatment, to convert

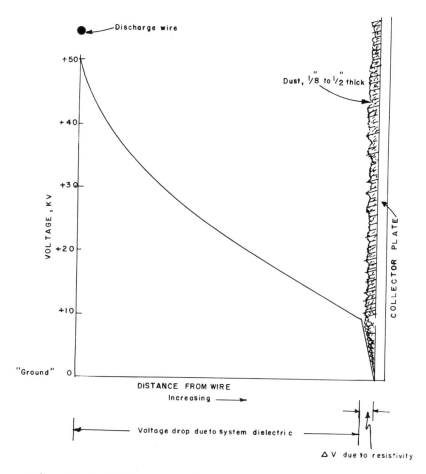

Fig. 9.9   Influence of Collected Dust on Electrostatic Precipitator Voltage Drop

dust that has either high or low resistivity to
an effective resistivity in the intermediate
range.  This is called dust conditioning.  Some
procedures utilized in actual practice to *reduce*
resistivity are:  spray the collected dust with
water; use water plus wetting agents; add sulfurous
acid to basic dusts; add ammonia to acids; and
add conductive materials such as triethylamine and
acid salts.

If it is not possible to condition the dust to stop back ionization, it then becomes necessary to lower the current density. The lower limit to which current can be decreased is about $2x10^{-2}$ amp/cm² measured at the discharge electrode surface. If the current is decreased lower than this, there will not be sufficient electrostatic charge available to ionize the gas and precipitate the dust.

It is important to recognize the fact that resistivity varies with temperature. Most materials exhibit an increase in resistivity as temperature is increased up to a critical temperature. Above this, the resistivity decreases with continued increase in temperature. This is shown in Figure 9.10 for cement dust and lead fume with percent moisture as the parameters. As would be expected, the resistivity decreases at any given temperature with increased moisture content. The resistivity of coal fly ash is similar to that of cement dust if 0.13 to 2.0% sulfur in coal is substituted for the 1.3 to 2.0% moisture parameter.

Resistivity can be calculated using an Arrhenius type equation:

$$R_b = A_R \exp(-E'/KT) \qquad (9.40)$$

where: 
- $R_b$ = bulk resistivity, ohm cm
- $A_R$ = resistivity constant of material, ohm-cm.
- E' = activation energy required for conduction, joules
- $K$ = Boltzman's constant = $1.38x10^{-23}$ joules/°K
- T = absolute temp., °K

Equation 9.40, as written, can be used to estimate resistivities for temperatures up to nearly the critical temperature. For temperatures slightly greater than the critical temperature, the sign of the exponent of e (natural logarithm base) is changed to positive. Values of the resistivity constant and activation energy can be obtained from experimental data.

## 9.6 THERMAL PRECIPITATION

It is possible to deposit particulate matter from a gas stream on a cold surface by thermal precipitation. It was stated earlier that aerosols

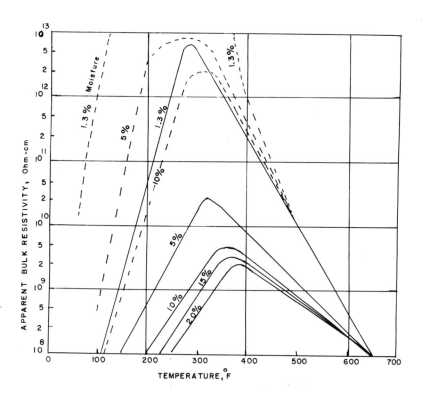

Fig. 9.]0    Effect of Moisture and Temperature on
             Resistivity (For Electrostatic
             Precipitation)

are repelled by heat and attracted by cold. This
is explained by the fact that the particles are
bombarded on the hot side by gas molecules which
are traveling at a higher average velocity (see
Equation 8.20) and, therefore, have a higher
kinetic energy. Molecules on the cold side of
the aerosol particles move at lower velocities and
push against the particle with a lower kinetic
force. As a result, the aerosol migrates toward
a cold area. Byers (5) shows that particle deposition by thermal force is a function of particle
size and aerosol temperature (which, in turn,

determines the thermal gradient). The highest
collection efficiency expressed as a fraction is
0.30 and is obtained for small particles ≤ 0.3
microns in diameter. This is a low collection
efficiency compared with other methods of
collection, but, if it could be improved, it would
be promising for this size particle. The efficiency
drops rapidly to about 0.05 as the particle
diameter increases to 1.2 microns. These efficiencies
can be increased using higher thermal gradients,
but the expenditure of large amounts of thermal
energy is not practical at this time because of
the low efficiencies involved.

Byers (5) suggests the use of the following
equations to calculate thermal force for various
size ranges according to Knudsen numbers ($\lambda$/r) as
follows:

*Range*          *Thermal Force Expression*   *Knudsen No.*

Free molecular $$F_t = \frac{-32\ r^2 C}{15\ v_g}\frac{dT}{dX}$$ >>1
regime

$$(9.41)$$

Continuum $$F_t = -9\pi r\frac{\mu^2}{\rho T}\frac{k_g}{2k_g+k_p}\frac{dT}{dX}$$ <<1
regime

$$(9.42)$$

    where:  $F_t$ = thermal force, $lb_f$
            C = translational part of gas thermal
               conductivity, Btu/(hr ft°F)
            r = radius of particle, ft
           $\mu$ = viscosity of gas, lb/(hr ft)
    $k_g$ & $k_p$ = thermal conductivity gas & particle,
               Btu/(hr ft°F)
           $\rho$ = density of gas, $lb/ft^3$
           T = absolute temp., °R
           X = distance, ft
          $v_g$ = velocity of gas molecule, ft/hr
          $\lambda$'= mean free path of molecules, ft
               (air ≈ $3.25 \times 10^{-7}$ft or $9.9 \times 10^{-8}$ m
               at SC)

It is also noted that Equation 9.42 is not applicable
for particles that have high thermal conductivity.

## 9.7 ATOMIZATION

Atomization of liquids plays one of the most useful roles in air pollution control. Atomized droplets of liquids are used as targets for inertial type impaction devices for removing particulate matter and, in addition, they serve as absorbers for removing soluble gaseous pollutants. Liquids can be atomized by at least five different procedures: 1) spray nozzles, 2) pneumatic two-fluid atomization, 3) spinning disks or cups, 4) impingement and 5) energy coupling. The names of the various atomization procedures simply indicates the physical mechanism by which the liquid is broken up.

Spray nozzles are designed to mechanically fracture the liquid by forcing it through a small nozzle under pressure. Some commercial nozzles are constructed to provide a centrifugal or swirl type of motion to the liquid which when added to the pressure and flow forces results in a greater total fracturing force. The liquid leaving the nozzle flies apart as it expands into the low pressure region and the droplet direction depends on the configuration of the nozzle. Figure 9.11A shows a full cone spray nozzle containing internal offset, removable vanes for centrifugal swirl. This type of unit is recommended for gas scrubbing devices. The nozzles in Figures 9.11B and C produce flat spray patterns with the first spray being projected straight foreward (in-line) and the second spray deflected at angles from 17-84°. The B and C sprays are useful for washing down and keeping the surfaces of collectors free from dust build-up

Pneumatic two-fluid atomization is accomplished by contacting one fluid stream (usually the liquid) with a high speed jet of a second fluid (usually a gas). The kinetic energy of the high speed stream is sufficient to overcome the surface force (surface tension) of the slow moving liquid causing the liquid to fracture. The result is that finely divided droplets of liquid can be obtained in the gas stream by this mechanism. This type of atom-ization differs from the other types in that the material to be removed by the atomized droplets is contained in the atomizing fluid. It should be remembered that both the targets (liquid drops) and the pollutants to be captured (particulates or gases in the gas stream) are definitely in motion with this type of atomization (they may or may not

LIQUID

LIQUID

(A) Wide angle full cone spray nozzle with
internal vanes

(B) In-line flat spray nozzle

LIQUID

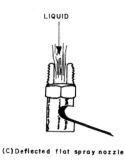

(C)Deflected flat spray nozzle

Figure 9.11   Several Types of Atomizing Spray
Nozzles (Spraying Systems Company Type)

be in motion in other types of atomization).  The
initial relative velocity differences are large,
however, this quickly decreases to zero as the
targets are accelerated to the gas velocity.  For
the collection of both particulates and gases by
impaction, it is desirable to have a high relative
velocity difference.  Two-fluid atomization can
be carried out in spray nozzles as shown in Figure
9.12A and B to combine advantages ob both procedures.
These types nozzles are frequently used in oil burners
as discussed in Section 8.5.2; but they are not
recommended when the gas stream contains much
particulate matter because the nozzles will plug.
Liquid and gas are both supplied under pressure to

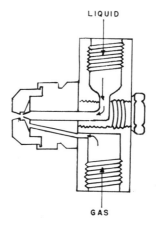

(A) PRESSURE PNEUMATIC SPRAY NOZZLE WITH INTERNAL MIXING

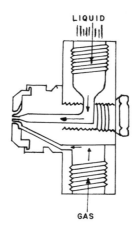

(ь) SIPHON PNEUMATIC SPRAY NOZZLE WITH EXTERNAL MIXING

Figure 9.12   Some Pneumatic Atomization Nozzles
                (Spraying Systems Company Type)

the internal mixing nozzle in Figure 9.12A while
the liquid is siphoned by an aspiration effect
from the external mixing nozzle in Figure 9.12B.
Figure 9.13 shows throat-spray type of pneumatic
atomizers of the type commonly used in Venturi
xcrubbers and other pneumatic atomizing collectors.
These systems handle gases containing either or
both particulate and gaseous pollutants without
plugging.

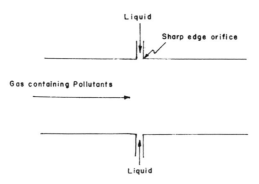

(A) Liquid enters under pressure from evenly spaced orifices

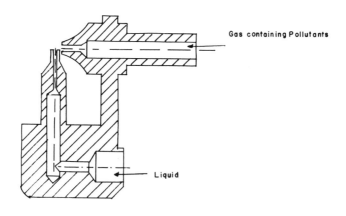

(B) Liquid enters by suction

Fig. 9.13   Throat Type Pneumatic Atomizers

Spinning disks, cups, arms or other related devices are used to physically fracture liquid by discharging it at high velocities from the periphery of the rapidly rotating device. The disks are usually flat or curved plates and either solid or perforated. Rotating arms are not truly spinning disks, but are a combination of nozzles (drilled holes) in pipes that are mechanically rotated to help break up and distribute the spray. The centrifugal force developed by a spinning disk

is sufficient to overcome the surface tension of
the liquid flowing across the disk  causing it to
break up into small droplets which are flung from
the edges of the disk in a 360° arc.

Impingement is sometimes used to produce
atomization by impinging two liquid jets upon
each other or by forcing a single jet to impinge
on a solid surface. Atomization by energy coupling
is the newest type of atomization but the mechanism
is not as well defined as it could be. The most
common types of energy used for this are sonic
(supersonic and subsonic) vibrations and high
voltage electrical energy. The most complete
source of information on sonic atomization is
presented in a translation of a Russian paper by
Mednikov (6). Sonic atomization occurs when a
high pressure resonating gas wave contacts a
liquid.

## 9.7.1  DROPLET SIZE PREDICTION

It is possible to predict the size of the
droplets created by atomization. Empirical data
available in manufacturer's literature is most useful
for predicting the size of droplets produced from
specific types of spray nozzles and rotating
devices. These data are usually easily available
in tabular form showing spray patterns, spray
coverage, quantity of liquid consumed and droplet
size for specific nozzles, liquid pressures and gas
flows and pressures when used.

At least two different types of atomization can
be made to occur by the pneumatic two-fluid
atomization procedure. The first, called drop-type
atomization, results when the liquid is introduced
into the high velocity gas stream from small
diameter (<1mm ID) inlet nozzles, or by introducing
the liquid into the gas stream in drop form (for
example, as from spray nozzles). Cloud-type
atomization is the second type and results when the
liquid is introduced as a stream (usually from
nozzles that are >1mm ID). Cloud-type atomization
results in the formation of much smaller droplets,
all of which appear to be less than 30 microns in
diameter. Cloud-type atomization was only discovered
  1968 by Hesketh (7).

Drop-type atomization has been studied
extensively for many years. The most popular and
thoroughly researched equation for predicting
average droplet size produced by drop-type

atomization is the Nukiyama-Tanasawa equation (8)
which is:

$$\bar{d}_x = \frac{585}{v'_g - v'_1}\sqrt{\frac{T}{\rho_1}} + \left(\frac{\mu_1}{T\rho_1}\right)^{0.45}\left(\frac{1,000\ Q_1}{Q_g}\right)^{1.5} \quad (9.43)$$

where:  $\bar{d}_x$ = Sauter mean drop diameter (see
                Table 7.1), microns
        $v'_g$ = velocity of gas, m/sec
        $v'_1$ = velocity of liquid, m/sec
        $\rho_1$ = density of liquid, g/cm$^3$
        $\mu_1$ = viscosity of liquid, g/(cm sec)
        $Q_1$ = liquid volumetric flow rate, m$^3$/sec
        $Q_g$ = gas volumetric flow rate, m$^3$/sec
        $T$ = surface tension, erg/cm$^2$

This equation simplified for air-water systems near
normal temperatures and pressures and when liquid
velocities are low, is given in English units as:

$$\bar{d}_x = \frac{16,400}{v_a} + 1.45\ (L')^{1.5} \quad (9.44)$$

where:  $v_a$ = velocity of air, ft/sec
        $L'$ = ratio of liquid to gas flow rate,
                gal/1000 ft$^3$

As an example, the Sauter mean drop diameter of
drop-type atomized water predicted by this equation
is 83$\mu$ when the air velocity is 245 ft/sec and the
liquid to gas ratio is 5 gal/1000 ft$^3$.
    Under the same conditions, if cloud-type
atomization had been used (larger inlet nozzles),
the mean diameter is estimated to be about 10
microns. It can be seen by Equation 9.44 that
*both* increased air velocity and decreased quantity
of liquid produce smaller atomized droplets.
    The very small droplets formed by cloud-type
atomization join together by hydrostatic force
without coalescing to form clouds that move
a single system which then has a much larger
effective diameter. Using atomized droplet, size,
velocity and acceleration observations, Hesketh
(9) shows that the <10$\mu$ droplets form clouds with
effective diameters from 170$\mu$ when the atomizing
air velocity is 150 ft/sec to 500$\mu$ when the air
velocity is 290 ft/sec.

## 9.7.2 GAS VELOCITY AND LIQUID NOZZLE ID

Critical minimum velocities of the gas stream are required for satisfactory pneumatic atomization. Hesketh (9) presents the following equation for predicting critical velocities for cloud-type atomization:

$$v_{g \ (crit)} = 1.7\left[\left(\frac{8,550}{d_n}\right)^{1/2} + 15.3\right] \quad (9.45)$$

where: $v_{g \ (crit)}$ = critical minimum velocity of gas stream, ft/sec
$d_n$ = liquid inlet nozzle ID, mm

This equation is good for nozzles with an ID of approximately 1mm or larger, because below 1 mm ID drop-type atomization occurs. This equation cannot be extrapolated for the drop-type atomization critical velocities. Velocities below 150 ft/sec are not recommended for gas stream atomization which is the reason for the 1.7 multiplier in Equation 9.45. 150 ft/sec is about the critical minimum velocity for both types of atomization at all nozzle diameters. The last term in the empirical Equation 9.45 is obtained from the fact that liquid streams are moving 15.3 ft/sec at the time when most of the atomization occurs.

Pneumatic atomization cannot be effective if the liquid introduced is not projected into the gas stream, and then it will be atomized only when the gas velocities are above critical. If the liquid nozzles are *too* large, the liquid is pushed against the walls and lost. The recommended maximum size liquid inlets should result in a 9:1 scrubber throat diameter (or height) to liquid nozzle diameter (see Figure 9.13A). The number of nozzles required should then be evenly spaced around the throat and directed toward the center of the throat.

### 9.7.3 ATOMIZATION EFFICIENCY

Very little of the available kinetic energy in high velocity gas streams is actually utilized for atomization. Hesketh calculates that the atomization efficiency based on work to form new surface area is approximately 6.1% for cloud-type atomization of water by air. Marshall (10) shows that typical drop-type or pressure nozzle atomization efficiency is 0.53% for water and approximately 0.17% for organic liquids with low surface tension (about 1/3 that of water) regardless of the liquid viscosity. These efficiencies are calculated directly from the diameters of the droplets produced and indicate only the intensiveness of the atomization. The extent of atomization is unknown for both types so it is generally assumed that all the liquid is atomized (which is not true).

Pneumatic atomization may consume several times the energy required for spray nozzle or spinning disk atomization depending on amount and pressure of the gas used. However, pneumatic atomization is the only method that can handle large quantities of liquid commercially and produce droplets with a small diameter (less than 15μ). Marshall (10) provides the following equations for estimating atomization power requirements: (a) When pneumatic atomization using an air-water system is utilized, the horsepower estimated at SC is:

$$Hp_g = 0.887 \; W_a \; \ln \frac{p_1}{p_2} \qquad (9.46)$$

where:  $W_a$ = wt. air used, lb/min
$p_1$ = initial gas pressure
$p_2$ = final gas pressure

(b) Pressure nozzle atomization power is estimated to be:

$$Hp_n = \frac{W_L(p_1-p_2)}{229 \; \rho_L} \qquad (9.47)$$

where:  $p_1$ = initial liquid pressure, psi
$p_2$ = final liquid pressure, psi
$W_L$ = wt. liquid atomized, lb/min
$\rho L$ = liquid density, lb/ft$^3$

The degree of atomization is not included in these theoretical equations which means that there will be some increase in power required over that theoretically estimated to produce the same degree of atomization at high liquid capacities.

Most spinning type atomizers utilize small amounts of air or gas to obtain more effective atomization. Green (11) shows that droplets as small as 6 microns in diameter can be produced using 50 psig gas pressure if the liquid flow rate is kept below 1 cm³/min. This would usually require about 125 liters per minute of air. The same study shows that, by reducing gas pressure to one pound per square inch, the average droplet diameter increases to about 100 microns , and a much higher liquid flow rate can be used. This requires 11 liters of gas per minute to make the 100 micron droplets.

Atomization is possible by practically any mechanism by which energy can be added to the liquid resulting in a breaking of the surface bonds. In addition to mechanical splashing devices, ultrasonics and electrical atomization are used but are currently of less importance in air pollution control. Ultrasonics can accomplish atomization by subjecting the main body of the liquid to a high frequency energy wave. Sound waves are longitudinal and, therefore, can be effective in producing atomization of a liquid whereas radio, light , and other transverse types of waves are not. Application of voltage potentials to freely falling drops can create an electrostatic repulsion force that can cause drops to burst into smaller droplets. Atomization efficiencies of these devices have not been adequately established.

QUESTIONS FOR DISCUSSION

1. Describe briefly some of the various methods available for removing particulate matter from a gas stream.
2. How would you determine if a particle can be considered spherical?
3. What influences whether or not a particle will adhere to a target?
4. How can densities of aerosols in the atmosphere be determined?
5. Where does the Boltzman constant come from?
6. Why are fog droplets usually greater than 5 microns in diameter? How is this value determined?
7. What is the difference between $N_{Re}$ and Re? Give an example of each.
8. What is the significance of coefficient of drag?
9. What is the value of the thermal conductivity of air? What is the value of the translational part of this thermal conductivity?
10. The dielectric constant of $CO_2$ for both radio and microwave frequencies is reported as:

$$(\varepsilon - 1) \; _X \; 10^6 = 921$$

What is the dielectric of $CO_2$?

PROBLEMS

9.1   A particle with a diameter of 15 microns and
a density of 1 g/cc is in free fall in air at
standard conditions.  Calculate the coefficient of
drag, the gravitational force and the terminal
settling velocity of this particle.
9.2   Calculate $v_S$ using a particle with a diameter
of 0.5 microns.
9.3   Repeat Problem 9.2 using a particle with a
diameter of 150 microns.
9.4   Repeat Problem 9.2 using a particle with a
cubical geometrical shape and a diameter of 30
microns.
9.5   If silica flour can be considered to have a
$d_{50}$ by number of 25μ and a σ of 1.4, devise a
system that would enable you to separate this
sample into groups having approximately a 10
micron diameter spread for each group using air
elutration.
9.6   Calculate the effective target efficiency
fraction for an inertial impactor where the target
is 150 micron water drops moving at an average
velocity of 5 cm/sec and the material being
collected is glass beads with diameters ranging
from 10 to 50 microns in a 30 cm/sec air stream.
9.7   What is the stopping distance of 1, 3 and 9
micron diameter coal fly ash entering an expansion
chamber in a gas stream that is initially moving
at 200 cm/sec?  If a baffle plate 6 centimeters
square is located 15 centimeters in front of the
particle as it initially enters, will the particle
strike the baffle?
9.8   A cyclone separator has an inlet opening four
inches wide by 10 inches long.  The collector is
2 feet in diameter.  What must be the minimum inlet
gas velocity to provide a separation factor of at
least 350 for all coal fly ash particles greater
than 15 microns in diameter?  What will be the
value of the normal velocity?
9.9   What is the diffusivity of 0.8 micron diameter
cement dust and 0.8 micron diameter lead dioxide
in air?

9.10  What is the fractional diffusional collection
efficiency for collecting 0.3 micron diameter pollen
on the lining of the upper respiratory tract?
(Consider the normal breathing capacity to be 125
liters/min, the trachea to be 25 millimeters I.D.
and respiration frequency is 1 per minute.)
9.11  What voltage is necessary to produce a
corona discharge in a cylindrical electrostatic
collector using 12 gauge (A.W.G.) discharge
electrodes and 3 inch ID collecting tube?  The
smoothness factor may be considered to be 0.95.
9.12  What change in the required corona voltage
occurs if the collector in Problem 9.12 used 8
gauge discharge electrode wire?  What is the
collection efficiency of this unit on 0.1 micron
particles in air if gas flow is 12 ft/sec?
9.13  Calculate the activation energy required for
conduction and the resistivity coefficient of pre-
cipitated cement dust at 10% moisture and at 200°
and 450°F.
9.14  If the ionization in the collector of
problem 9.11 occurs in the first 0.5 meters and if
it is operating at a voltage potential of 50 KV,
what thickness of cement dust with 10% moisture
could be permitted to build up at 200°F before
back ionization would prevent effective collection?
9.15  What is the effective collection efficiency
on cement dust with an average diameter of 12
microns and containing 10% moisture for the
collector of 9.11 operating at 40 KV with 1/16
inch dust buildup at 200°F if the gas velocity is
kept at the lower portion of the critical velocity
for cement and if the collector is 30 cm long?
9.16  What is the thermal force present on a 6
micron diameter cement particle as it approaches
the wall of the precipitator of 9.11 if the wall
temperature is 80°F and the gas temperature is
200°F?  How does this compare with the electric
force on the same particle?

REFERENCES

1.  Johnstone, J.F. and M.H. Roberts, "Deposition
    of Aerosol Particles from Moving Gas Streams,"
    I & EC, Vol. 41, No. 11, pp. 2417-2423 (1949)
2.  Spiegel, M.R., "Theory & Problems of Vector
    Analysis," Shaum's Outline Series, Shaum
    Publishing Co., (1959)
3.  Peek, F.W., Jr., "Dielectric Phenomena in
    High Voltage Engineering," 3rd Edition, McGraw-
    Hill, N. Y. (1929)

4.  Gottschlich, C.F., "Source Control by
    Electrostatic Precipitators," Chapter 45 of
    Air Pollution, 2nd Edition, Edited by A.C.
    Stern (1968)
5.  Byers, R.L., and S. Calvert, "Particle
    Deposition from Turbulent Streams by Means of
    Thermal Force," I & EC Fundamentals, Vol.8,
    pp. 646-655 (1969)
6.  Mednikov, Evenii Pavlovich, "Accoustical
    Coagulation and Precipitation of Aerosols,"
    Plenum Publishing Corp., N. Y. (1965)
7.  Hesketh, H.E., A.J. Engel, and S. Calvert,
    "Atomization--a New Type for Better Gas
    Scrubbing," Atmospheric Environment, Vol. 4,
    No. 7, pp. 639-650 (1970)
8.  Nukiyama, S. and Y. Tanasawa, "An Experiment
    on the Atomization of Liquid by Means of an
    Air Stream," Trans. Soc. Mech. Eng. (Japan),
    Vol. 4, No. 14, p.86 (1938)
9.  Hesketh, H.E., "Gas Cleaning Using Venturi
    Scrubbing", paper 72-111, 65th Annual Air
    Pollution Control Association meeting, Miami,
    Florida (June 19, 1972)
10. Marshall, W.R., Jr., "Atomization and Spray
    Drying," Chem. Eng. Progress, Monograph Series,
    No. 2, Vol. 50 (1954)
11. Green, H.L. and W.R. Lane, "Particulate Clouds-
    Dusts, Smokes & Mists," Chapter 2, Second
    Edition, Universities Press (1964)

# CHAPTER X

# GASEOUS POLLUTANT REMOVAL THEORY

Pollutants in the form of gases and vapors usually make up only a small portion of a waste gas stream. Quite frequently, nitrogen (from the air) comprises the largest single component of waste gas streams. This means that the total amount of gaseous pollutants to be removed is small but the low concentration materials are usually the most difficult to remove. Gaseous pollutants can never be totally removed from a gas stream. However, it is possiblt to effectively reduce the concentration of *any* gaseous pollutant to an acceptable level if cost were not a factor.

Most of the procedures for removing gaseous pollutants are standard chemical engineering operations with the primary difference being that normal chemical engineering operations deal with higher concentrations. This chapter on the removal of gaseous pollutants should provide the reader with a general familiarity of gas purification theory. The dilution effect of the atmosphere by atmospheric diffusion is discussed in Chapter III. Also, atomization of liquids necessary for gas absorption is included in Chapter IX.

## 10.1 DIFFUSION OF GASES

Diffusion of gases depends equally on the nature of the fluids as well as physical conditions. For example, gas density, viscosity and temperature are all important, but so is the turbulence of the system. Atmospheric eddy diffusion depends almost entirely on conditions which cause mechanical disturbances and on such factors as temperature and pressure gradients. Atmospheric diffusion, as pointed out in Chapter III, neglets molecular diffusion and utilizes only the system turbulence.

Man-made chemical operations (mini-sized compared with atmospheric systems) utilize both eddy diffusion, which is created by the system turbulence and flow, as well as molecular diffusion to achieve good mass transfer.  The turbulence induced eddy diffusion brings the molecules in the main body of the fluid to a position near the phase boundary.  From here, the slower molecular diffusion takes over and moves the molecules through the boundary.  Molecular diffusion can also have the important function of moving molecules through stationary surfaces or layers of the fluid which have no eddy motion because of the viscous drag.

The rate of the lineal diffusion of fluid molecules can be estimated by the following equations (gaseous fluids will be treated before liquids). For the diffusion of gases at near atmospheric conditions, the diffusivity can be calculated using a modified Gilliland (1) type of equation:

$$D'_{AB} = 0.0325 \frac{(T)^{\frac{1}{2}}}{(\tilde{V}_A^{1/2} + \tilde{V}_B^{1/2})^2} \left[ \frac{1}{M_A} + \frac{1}{M_B} \right]^{\frac{1}{2}}$$

$$(10.1)$$

where:  $D'_{AB}$ = gas diffusivity, lb moles/(ft hr)
$T$ = absolute temp, °R
$M$ = molecular weight
$\tilde{V}$ = molecular volume of the gas when a liquid at the normal boiling point, cm$^3$/(g mole)

The diffusivity of gas A into gas B ($D_{AB}$) is equal to the diffusivity of B into A ($D_{BA}$).  Table 10.1 lists some values of $\tilde{V}$.

Diffusion in liquids can be estimated from the Wilkie Equation which is a modification of the Stokes-Einstein Equation:

$$D_{AB} = 7.4 \times 10^{-8} \frac{(\Psi_B M_B)^{1/2} T}{\mu \tilde{V}_A^{0.6}} \qquad (10.2)$$

where:  A = solute (material dissolved)
B = solvent (dissolving material)
$D_{AB}$ = Liquid diffusivity, cm$^2$/sec
$\mu$ = solution viscosity, centipoise
$T$ = absolute temp., °K
$\tilde{V}_A$ = molar volume of solute as liquid at the normal b.p., cm$^3$/(g mole)

$\Psi_B$ = a constant for the solvent
= 2.6 for $H_2O$, 1.9 for methanol, 1.5 for ethanol, and 1.0 for benzene, ether, heptane and similar solvents

Equation 10.2 has the diffusivity expressed in metric units of area per time. This same equation expressed in English units of mass per distance per time for dilute ideal liquids is:

$$D'_{AB} = 1.6 \times 10^{-7} \left(\frac{\Psi_B}{M_B}\right)^{1/2} \frac{\rho_T}{\mu \, \tilde{V}_A^{0.6}} \quad (10.2a)$$

where:  $D'_{AB}$ = liquid diffusivity, lb moles/ (ft hr)
$\rho$ = solution density, $lb/ft^3$
$T$ = absolute temperature, °R

TABLE 10.1  MOLECULAR VOLUMES OF SOME GASES IN A LIQUID STATE AT THE NORMAL BOILING POINT

| Substance | Molecular Volumes cc/(g mole) |
|---|---|
| Air | 29.9 |
| $H_2$ | 14.3 |
| $CO_2$ | 34.0 |
| CO | 30.7 |
| $O_2$ | 14.8 |
| $N_2$ | 36.0 |
| NO | 23.9 |
| $NO_2$ | 32.2 |
| $SO_2$ | 40.4 |
| $SO_3$ | 47.8 |
| $C_2H_6$ | 51.8 |
| $C_6H_5COCl$ | 136.1 |
| $H_2O$ | 18.8 |

## 10.2  MASS TRANSFER AND TWO-FILM THEORY

In order to transfer a given quantity (mass) of gaseous pollutant from a gas to a liquid, it is necessary to transfer the molecules through a boundary region called the "film". Molecules travel in both directions through the film by diffusion, some coming from the gas side and some coming from the liquid side. The net rate of transfer will be influenced by the driving forces such as temperature, pressure, concentration and solubility. Concentration of molecules on the gas side is affected by both

eddy and molecular diffusion. Normally, eddy
diffusion is established to maintain a uniform
concentration of molecules in the bulk of the gas
phase. Molecular gaseous diffusion then provides
the means by which the gas molecules move to the
film. On the liquid side, a similar situation
exists where eddy diffusion can be created to
maintain essentially a constant concentration in
the bulk of the liquid and liquid diffusion moves
molecules to and from the film. If the molecules
of gas that pass through the film and into the
liquid chemically react or are otherwise physically
removed to keep them from passing back through the
film, the net rate of transfer from the gas to
the liquid is increased.

It is convenient to consider the boundary
that exists between the gas and the liquid phases
as being composed of two separate films--one being
the gas film and the other the liquid film. This
is known as the two-film theory and is shown
pictorially in Figure 10.1.

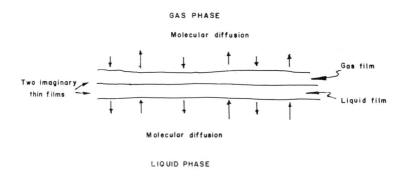

Figure 10.1   Molecular Transfer Through Two-Film
              Boundry

The net overall resistance is considered to be due
to the combined gas and liquid films and is equal
to the reciprocal of the sum of the reciprocal of
the individual resistances. Usually, the resistance
of one of the films is significantly larger and it
is convenient to attribute the total resistance to
the passage of molecules through the film to the
film with the largest resistance. For example, if

the molecules can pass rapidly through the gas side
of the film, then the net resistance to the passage
of molecules is attributed to the liquid film and
this is called "liquid phase controlling". If most
of the resistance is in the gas film, then it is
called "gas phase controlling" and the resistance
due to the liquid side is neglected.

The rate at which mass is transferred from one
phase to another is proportional to the transfer
rate constant called a mass transfer coefficient
(k). Separate equations can be written to show these
coefficients for each film. The equation for the
gas film is:

$$N_A = k_G A(p_{AG} - p_{Ai}) \qquad (10.3)$$

For the liquid film, the equation is:

$$N_A = k_L A(c_{Ai} - c_{AL}) \qquad (10.4)$$

where:  $N_A$ = rate of transfer of A through the
          film, lb moles/hr

$k_G$ = gas film coefficient, lb moles/(hr
          ft$^2$ atm)

$k_L$ = liquid film coefficient, lb moles/
          (hr ft$^2$ lb moles/ft$^3$)

$A$ = surface area of mass transfer, ft$^2$

$p_{AG}$ = partial pressure of A in gas phase,
          atm

$p_{Ai}$ = partial pressure of A at vapor--liquid
          interface, atm

$c_{Ai}$ = concentration of A at vapor--liquid
          interface, lb moles/ft$^3$

$c_{AL}$ = concentration of A in liquid phase,
          lb moles/ft$^3$

The individual mass transfer coefficients ($k_G$ & $k_L$)
can be obtained from experimental data. It is
convenient to relate these individual mass transfer
coefficients to overall mass transfer coefficients
($K_G$ & $K_L$) which are assigned to either the gas or
the liquid phase. This is done as discussed in
Section 10.3.

## 10.3  GAS ABSORPTION

Gas absorption is the taking in of a gas or
vapor by a liquid and includes both physical and
chemical absorption. In absorption, the material
taken up is distributed throughout the entire

absorbent (liquid) phase. Absorption is distinguished
from adsorption in that the latter is a surface
phenomena with the material taken up being distributed
over the surface of the adsorbing material.
In order to be absorbed, pollutant molecules
must diffuse to the gas film, pass through both the
gas and the liquid films and finally diffuse into
the liquid. This means that the net transfer of
molecules through either film is the same, so we
can set Equations 10.3 and 10.4 equal to each other.

$$N_A = k_G \, A(p_{AG} - p_{Ai}) = k_L \, A(c_{Ai} - c_{AL}) \qquad (10.5)$$

It is not common to assign values of pressures
and concentrations at the interface, therefore
overall mass transfer coefficients are used with
pressure and concentration values that reflect the
overall or complete gradient across both films.
The overall mass transfer equations using overall
absorption coefficients can be written for both
films. At equilibrium conditions, we have:

$$N_A = K_G \, A(p_{AG} - p_{AL}) = K_L \, A(c_{AG} - c_{AL}) \qquad (10.6)$$

where: $N_A$ = rate of absorption of A, lb moles/hr
$\quad\quad\;\; A$ = surface area for absorption, $ft^2$
$\quad\quad\;\; K_G$ = overall gas absorption coefficient,

$\quad\quad\quad\quad$ lb mole/(hr $ft^2$ atm)

$\quad\quad\;\; K_L$ = overall liquid absorption coefficient,

$\quad\quad\quad\quad$ lb mole/(hr $ft^2$ lb mole/$ft^3$)

$\quad\quad\;\; p_{AL}$ = partial pressure of A if it were in
$\quad\quad\quad\quad$ equilibrium with a liquid solution
$\quad\quad\quad\quad$ having that concentration of A, atm

The $p_{AL}$ and the $c_{AG}$ terms are pseudo-quantities.
There is obviously not a partial pressure of A in
the liquid, however, $p_{AL}$ refers to a vapor condition
that could exist above the liquid. Note that no
defining statement is made for $c_{AG}$ which could be
thought of as being the concentration of A in the
gas in units of lb moles/$ft^3$. Equation 10.6 can
also be utilized for extraction processes where it
is desirable to know the rate of desorption
(extraction) of a dissolved gas from a liquid. In
this case, the signs would be reversed for all
terms in parentheses in both parts of the right
hand of Equation 10.6.

## 10.3.1  GAS LAWS

In order to obtain the quantity $p_{AL}$, it is necessary to combine some of the basic gas and solution laws. From the gas fundamentals, we know that Boyle's Law (the product of pressure and volume is a constant at constant temperature) and Charles' Law (pressure divided by absolute temperature is a constant at constant volume) result in what we know as the Ideal Gas Law which, for one mole of material, says that the product of pressure times volume divided by temperature is a constant. The pressure term in the equation just discussed can include contributions from one or more gases depending on whether the gas under consideration is pure or a mixture. Dalton's Law of partial pressures states that mixtures of non-reacting gases exhibit independant pressures, the sum of which is the total pressure of the system. Air is frequently considered to be essentially a binary system of non-reacting gases (nitrogen and oxygen). For such a system consisting of components A and B, Dalton's Law becomes:

$$P_{AG} + P_{BG} = P_T \qquad (10.7)$$

where:  $P_T$ = total gas pressure

If these gases can be considered to be ideal, we can substitute the ideal gas law expression either for individual components:

$$P_{AG}V = N_{AG}RT \qquad (10.8)$$

where:  N = number of moles of gases
        V = gas volume
        R = universal gas constant
or with Dalton's Law also for the system:

$$(P_{AG} + P_{BG})V = (n_{AG} = N_{BG})RT \qquad (10.8a)$$

Dividing Equation 10.8 by Equation 10.8a shows that for each component in an ideal gas system pressure ratio equals mole ratio. Amagat's law states that the total volume of an ideal, non-reacting gaseous system is equal to the sum of the individual pure component volumes. Combining Amagat's law and the ideal gas law shows that volume ratio equals mole ratio for each component. These relationships are extremely important and must be remembered:  volume most ratio = pressure ratio = mole ratio for ideal and and most dilute gas systems (i.e., $V_{AG}/VT = p_{AG}/P_T = N_{AG}/N_T$).

The most common way of writing the relationship of Dalton's Law and the ideal gas law is:

$$P_{AG} = y_A \, P_T \qquad (10.9)$$

where: $y_A$ = mole fraction A in the gas

$$= \frac{\text{moles A}}{\text{moles total gas}}$$

## 10.3.2   SOLUTION LAWS

The behavior of solutions can be predicted by Raoult's and Henry's laws.   Raoult's Law is for *concentrated* ideal solutions where the components do not interact.   At equilibrium conditions, this equation can be written as:

$$P_{AL} = x_A \, P_A^\circ \qquad (10.10)$$

where: $x_A$ = mole fraction of A in solution

$$= \frac{\text{moles A}}{\text{moles total liquid}}$$

$P_A^\circ$ = vapor pressure of *pure* A at the same temperature and pressure as the solution

Vapor pressures of pure substances at various temperatures and pressures are available in many of the numerous chemistry, physics and engineering handbooks (values for water can also be easily obtained from the steam tables).
A solution of 30% glycerine in water at 212°F is a solution which is concentrated in water and can be used to demonstrate Raoult's Law.   The mole fraction of water is:

$$x_A = \frac{70/18}{70/18 + 30/92} = 0.923$$

We know from experience that the vapor pressure of pure water at its boiling point is equal to the atmospheric pressure.   The resulting vapor (partial) pressure of water in equilibrium with this solution is:

$$P_{AL} = (0.923)(14.7) = 13.6 \text{ psia}$$

It so happens that the vapor pressure of glycerine at this temperature is essentially 0.   The total pressure then remains 13.6 psia, showing that this solution does not boil at 212°F when the total atmospheric pressure is 14.7 psia. The boiling point under these conditions can be estimated by determining the temperature that will cause water to have a vapor pressure of 14.7/0.923 psia.

Henry's Law is for *dilute* solutions and can be written at equilibrium conditions as:

$$p_{AL} = c_{AL} \, H \qquad (10.11)$$

where: $p_{AL}$ = equilibrium pp of A over soln, atm
$c_{AL}$ = conc. of A in liquid phase, lb moles/ $ft^3$
$H$ = Henry's Law constant, $\dfrac{atm \, ft^3}{lb \, mole \, A}$

Henry's Law constants can only be determined empirically--by experiment. This law is really just a special form of Raoult's Law and, in this text, is written as such. The significant difference between the two laws is that Henry's Law is for dilute solutions and Raoult's Law is for concentrated solutions. The reader is cautioned that Henry's Law is written in many different ways, two of the more frequent being $c_{AL} = p_{AL} \, H'$ and $p_{AL} = x_A \, H''$. Units of H should always be examined to determine the form of equation for which it is intended to be used.

The mass transfer coefficients can be related to each other at equilibrium conditions. When Henry's Law applies, appropriate quantities in Equations 10.5 and 10.6 can be substituted for, yielding the following relationships:

$$K_L = K_G \, H \qquad (10.12)$$

$$\frac{1}{K_G} = \frac{1}{k_G} + \frac{H}{k_L} \qquad (10.13)$$

$$\frac{1}{K_L} = \frac{1}{k_L} + \frac{1}{H k_G} \qquad (10.14)$$

The fractional total resistance to mass transfer due to the gas film then becomes:

$$\frac{\frac{1}{k_G}}{\frac{1}{K_G}} = \frac{K_L}{H k_G} \qquad (10.15)$$

and the corresponding equation for the fractional total resistance to mass transfer due to the liquid film is:

$$\frac{\frac{1}{k_L}}{\frac{1}{\overline{K}_L}} = \frac{HK_G}{k_L} \qquad (10.16)$$

The Henry's Law constants for some gases dissolved in water at near ambient conditions have a linear relationship with temperature. The Henry's Law constant for carbon monoxide dissolved in water can be found by:

$$H = 263t + 10,100 \qquad (10.17)$$

where:  $H$ = atm ft$^3$/lb mole CO
        $t$ = temperature, °C

Figure 10.2 shows Henry's Law constants for carbon monoxide as well as for several other common pollutants in water over the temperature range 0 to 35°C. Values of the constant for gases in organic liquids can be estimated using:

$$H = \frac{22,400}{L \, \tilde{V}_L N_T} \qquad (10.18)$$

where:  $\tilde{H}$ = atm ft$^3$/lb mole gas
        $\tilde{V}_L$ = molar volume of liquid at 20°C, cm$^3$/ g mole
        $N_T$ = total number lb moles per ft$^3$ liquid
        $L$ = Oswald coefficient, cm$^3$ gas/cm$^3$ liquid
          $\simeq \exp\left[-\frac{T}{40} + 0.1\right]$
where:  $T$ = surface tension, dynes/cm (see symbols table)

10.3.3   INTERFACIAL AREA AND AVERAGE PRESSURE DIFFERENCE

The contact surface area between two phases in an absorber could be an extremely difficult quantity to evaluate. In the case of a simple wetted wall column operating with a *smooth* continuous film, the contact area can be readily calculated. However, for cases such as spray columns, packed beds or even wetted wall columns with irregular liquid surface,

# HENRY'S LAW CONSTANT FOR GASES IN WATER

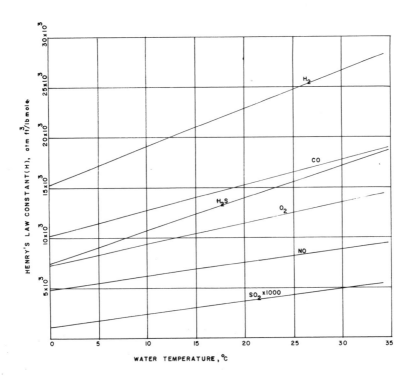

Figure 10.2  Henry's Law Constant for Gases in Water

calculation of the area can be very tedious and inaccurate.  For this reason, an interfacial area, which is a pseudo-quantity, is used.  It is:

$$a = \frac{A}{V} \qquad (10.19)$$

where:  a = interfacial area per unit of bulk volume
  A = actual contact area
  V = total volume of *empty* absorber

The interfacial area term is then combined with
the overall transfer coefficient and the product
of the coefficient times a is empirically determined.
Equation 10.12 can be rewritten as:

$$K_L a = K_G a \; H \qquad (10.20)$$

and Equation 10.6 can be rewritten using the interfacial area:

$$N_A = K_G a \; V(p_{AG} - p_{AL}) \qquad (10.21)$$

In the counter-current absorbers, the component
being absorbed from the gas phase has a decreasing
partial pressure in the gas phase and a decreasing
equilibrium partial pressure from the liquid phase
in contact with the gas as the gas passes through
the system and is absorbed. (Co-current absorbers
have the opposite relationship.) The equations
presented require point values so an average partial
pressure difference for either system is necessary.
For cases where significant changes occur, use a
log mean pressure difference as shown by the
following modification of Equation 10.21:

$$\Delta N_A = K_G a \; V(p_{AG} - p_{AL})_{ln} \qquad (10.22)$$

where:   $\Delta N_A$ = lb moles of absorbable material
              entering tower minus lb moles
              absorbable material leaving the
              tower per hour, all in the gas
              phase (or equivalent in terms of
              liquid phase).

$$(p_{AG} - p_{AL})_{ln} = \frac{(p_{AG} - p_{AL})_1 - (p_{AG} - p_{AL})_2}{\ln \dfrac{(p_{AG} - p_{AL})_1}{(p_{AG} - p_{AL})_2}}$$

The subscript ln signifies mean log mean difference
and is the natural log. The subscripts 1 and 2
used above refer to the pressures at the bottom
and pressures at the top respectively. (It makes
no difference which is called 1 or 2, but values
corresponding to each phase at the point they
contact whether it be top or bottom, must be
properly assigned.)

## 10.3.4 EQUILIBRIUM AND DRIVING FORCE

Equilibrium between a gas and a liquid is obtained when the rate of absorption of the gaseous component from the gas phase is equal to the rate of release of the absorbed gas from the liquid phase. Theoretically, equilibrium is only possible when the two phases have been in contact for an infinite period of time, but it is possible to approach nearly equilibrium conditions in a reasonable period of time for many substances. The vapor-liquid equilibrium constant ($K_i$) is expressed as follows:

$$K_i = \left(\frac{y_A}{x_A}\right)_i \qquad (10.23)$$

where:  i represents each individual specific concentration ratio at a given temperature and pressure
$x_A$ = mole fraction A in solution
$y_A$ = mole fraction A in gas

The value of $K_i$ changes with concentration and also changes with temperature and pressure. Equilibrium data can be used to construct equilibrium curves such as shown in Figure 10.3. The slope of this curve at any point is $K_i$.

Equilibrium curves can also be approximated using the generalized equations presented already. Dalton's Law for an ideal gas (Equation 10.9) and Raoult's Law for an ideal solution (Equation 10.10) are equal at equilibrium conditions:

$$y_A \, P_T = x_A \, P_A^o$$

and can be rearranged to give the equation of a line intersecting at the origin:

$$y_A = \frac{P_A^o}{P_T} \, x_A \qquad (10.24)$$

The ratio of the vapor pressure of the pure component at that temperature and pressure over the total pressure is sometimes equal to the slope of the equilibrium curve. However, this is only approximate and was derived for ideal gas and concentrated ideal solution conditions.

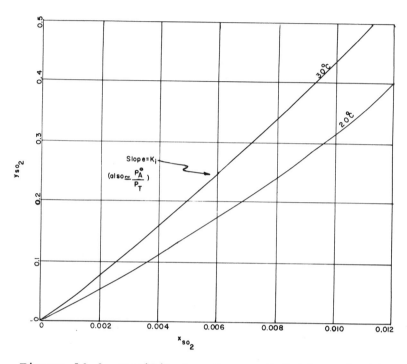

Figure 10.3  Equilibrium Curve of Sulfur Dioxide
in Water at 1 ATM

For the special case where Henry's Law applies
(Equation 10.11 for dilute ideal solutions), we can
obtain another approximate expression for the slope
of the equilibrium curve:

$$y_A = \frac{H}{P_T} c_A \qquad (10.25)$$

For a dilute binary solution, the mole fraction of
A can be expressed:

$$x_A = \frac{c_A}{c_A + c_B} \simeq \frac{c_A}{c_B}$$

because $c_A \ll 1$ and $c_B \simeq 1$. As a result, $x_A \simeq c_A$
and therefore, for the special case of a very dilute
ideal binary system:

$$y_A \simeq \frac{H}{P_T} x_A \qquad (10.25a)$$

In addition to the physical absorption of a gas into a liquid, there is also the possibility that a chemical reaction can occur. This is called chemisorption. If there is sufficient chemical reaction, it is no longer possible to consider the solutions ideal because of the interacting of molecules. For example, when $SO_2$ is absorbed by water, this is mainly a physical absorption mechanism with little chemical reaction. However, when $SO_2$ is absorbed in sodium hydroxide, there is an extensive chemical reaction (evidenced in this case by the release of heat of reaction). When chemisorption occurs, the component absorbed is no longer present to exert a partial pressure and the above equations cannot be used. It is quite useful at times to utilize both chemical reactions and diffusion to improve gas collection. This is discussed further in Section 10.3.7.

Counter-current type of absorbers are most frequently used in air pollution control. They can be schematically represented as shown in the drawing in Figure 10.4. For these types of devices, the scrubbing liquid enters the top of the system, is distributed evenly, then falls through the device by gravity. The gas stream containing pollutants to be absorbed enters at the bottom where it is distributed and then forced up through the system. Absorption occurs as the rising gas contacts the falling liquid. The cleaned gases leave from the top and the liquid containing the absorbed material leaves at the bottom. We can show the concentrations at the bottom of the absorber as $x_1$, $y_1$ in Figure 10.5 and conditions for the top of the absorber as $x_2$, $y_2$. The vertical distance from either of these points to the equilibrium curve represents the equilibrium driving force. The greater the distance, theoretically, the greater the driving force attempting to produce absorption. Concentrations of the absorbable material in the liquid and gas phases at various positions within the absorber can be represented by a line drawn between points $x_1$, $y_1$ and $x_2$, $y_2$ called the operating line (which may be straight or curved as discussed in the next section). Figure 10.5 shows an average driving force in the absorber as represented by the distance from the midpoint of the operating line to the equilibrium curve. If there is sufficient solubility of the gas pollutant being absorbed in the liquid, and if there is sufficient liquid to absorb the gas, it should be possible to reduce the

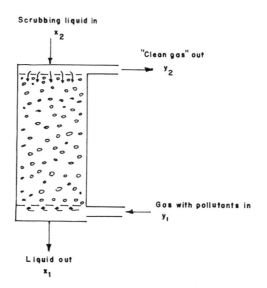

Scrubbing liquid in
$x_2$

"Clean gas" out
$y_2$

Gas with pollutants in
$y_1$

Liquid out
$x_1$

Figure 10.4   Counter-Current Absorber

concentration of the pollutant in the gas stream to the equilibrium concentration. This is possible only if the absorber permits adequate contact time (which may be too long a contact time).

10.3.5   ABSORBER OPERATING LINES

The operating line for a counter-current type absorber was described as a straight line in Section 10.3.4. It is frequent..y possible to use this linear relationship to depict actual gas and liquid concentrations inside the tower, but sometimes this approximation is not accurate. In order to understand how the operating line is derived, visualize a counter-current absorber as shown in Figure 10.6. L represents the total moles in the liquid phase in lb moles/(ft$^2$ hr) and G is the total moles in the gas phase in lb moles/(ft$^2$ hr). The subscripts ı and 2 indicate the bottom and top respectively of the absorber.

Assuming no accumulation and no chemical reaction, the total mass balance on the absorber at steady state amounts to the sum of materials

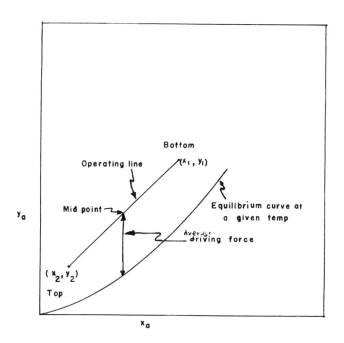

Figure 10.5 Equilibrium Curve and Driving Force
for Counter-Current Absorber

in equals the sum of materials out:

$$G_1 + L_2 = G_2 + L_1 \qquad (10.26)$$

The overall balance on the component A which is absorbed, can be written as:

$$G_1 y_1 + L_2 x_2 = G_2 y_2 + L_1 x_1 \qquad (10.27)$$

where: x and y = mole fraction A in liquid and
gas respectively.

If the amount of material absorbed from the gas phase is small, there will be no appreciable change in the total number of moles in the gas phase and $G_1$ can be assumed to equal $G_2$. Also, if only a small amount of material is absorbed and if no liquid volatilizes, then $L_1$ almost equals $L_2$. For these conditions, Equation 10.27 can be rewritten to express the slope of the operating line as:

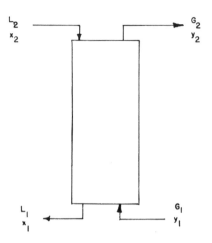

Figure 10.6   Overall Mass Balance on Counter-Current
Absorption Tower

Slope of operating line = $\dfrac{L}{G} = \dfrac{y_1 - y_2}{x_1 - x_2} = \dfrac{\Delta y}{\Delta x}$   (10.28)

Counter-current absorbing can be likened to
passing gas and liquid phases counter-currently
through a series of separate contacting stages as
shown in Figure 10.7. The subscripts in this
example refer to the stage from which each
component comes. In each of the $n$ theoretical
stages, there is good mixing and sufficient time
to permit equilibrium between the vapor and the
liquid to be obtained. Instead of having physically
separated stages with separation facilities and
connecting piping, put all of these stages together
in one continuous device, such as the counter-
current tower in Figure 10.6 which could be a
vertical cylindrical shell containing a number of
horizontally spaced plates. These plates, for
example, can be perforated to permit the gas to flow
upward through them, yet contain liquid on them
for contacting the gas. The top section of such
a column is shown in Figure 10.8. If we assume
equilibrium is reached on each plate, then these
are theoretical stages or plates. In actual
practice, equilibrium is not usually attained

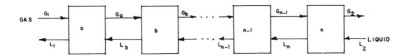

Figure 10.7  Counter-Current Contact Stages

because the absorber is not 100% efficient, so there
is a difference in the number of theoretical and
the number of actual stages.

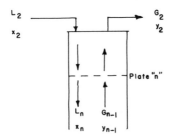

Figure 10.8   Mass Balance on Top Plate of Counter-
Current Absorber

A mass balance on the component absorbed can
be made around the top (nth) plate shown in Figure
10.8.  This produces the equation:

$$G_{n-1} \; y_{n-1} + L_2 \, x_2 = G_2 \, y_2 + L_n \, x_n \qquad (10.29)$$

When this is rearranged, we obtain the equation of
a line:

$$y_{n-1} = \frac{L_n}{G_{n-1}} \, x_n + \frac{G_2 y_2 - L_2 x_2}{G_{n-1}} \qquad (10.30)$$

Once again, this is the equation of the operating
line and may or may not be a straight line.  In
order to be a straight line with the slope L/G,
it is necessary that:  a) The concentration of
absorbable component be low so there would be no
appreciable change in L or G due to transfer of
this component; b) The solvent be relatively non-

volatile so there would be essentially no loss of
the liquid phase and c) The carrier gas which
comprises the bulk of the gas phase should not be
soluble in the liquid solvent.
     It is possible to operate absorbers using co-
current flow of the two fluids. Figure 10.9 shows
co-current flow with both the liquid and gas
entering at the bottom of an absorber. If there
is no appreciable transfer of mass between the gas
and liquid phases, the overall mass balance for
co-current absorbers then yields an operating line
with a negative slope:

$$\text{slope of operating line} = -\frac{L_M}{G_M} = \frac{y_1 - y_2}{x_1 - x_2} \qquad (10.31)$$

With these assumptions, the operating line would
be straight and oriented in a manner similar to
that shown in Figure 10.10 (when the column is top
fed, reverse the "Top-Bottom" notations).

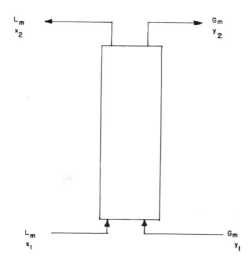

Figure 10.9   Overall Mass Balance for Co-Current
           Absorption Tower

     Cross-current flow in absorbers is also used
in air pollution control. An example of this is
a tower using intermediate spray injections.
Operating lines for these cases are more complex
and must be determined for the specific geometrical
configuration used.

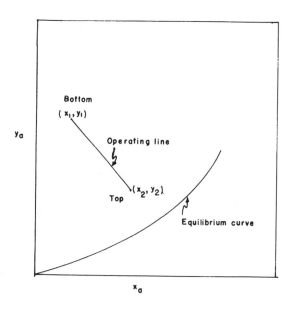

Figure 10.10   Co-Current Absorption Equilibrium
            Diagram and Operating Line

10.3.6   CONTACT STAGES AND EFFICIENCY

The existance of a difference between
theoretical (ideal) contact stages and actual
contact stages was noted in Section 10.3.5.
Attempts are made to design absorbers for maximum
efficiency by trying to optimize physical contact
mechanisms such as good mixing and high mass
transfer rate as well as by designing for maximum
contact time.  Even so, it is unusual to
allow equilibrium to be reached in actual practice
because this could require an infinite number of
contact stages.  The number of theoretical contact
stages can be estimated by "stepping off" the
contact points on an equilibrium diagram.  This is
done as shown in Figure 10.11 by drawing horizontal
and vertical lines connecting the operating and
equilibrium lines.  (In the example shown in Figure
10.11, the absorbing liquid entering the top of
this counter-current tower is pure and contains
no absorbable material.  For this reason, the

point $x_2$, $y_2$ is located on the ordinate.)  The
position on the equilibrium curve indicated as x'
is the theoretical equilibrium concentration of
liquid *leaving* the top plate.  At this same point,
the composition of the vapor (y') is the theoretical
equilibrium concentration of vapor which is rising
*from* the top plate.  The intersection of each
horizontal and vertical line on the equilibrium
curve indicates a similar situation for the
successive plates.  In the example given in Figure
10.11, there are approximately 2 1/4 theoretical
plates (ideal contact stages).

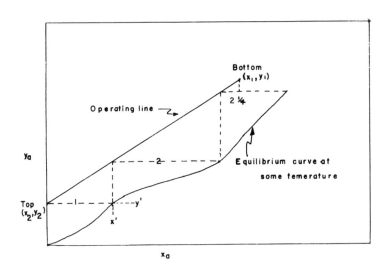

Figure 10.11    Ideal Contact Stages for Counter-
Current Absorber

Assuming that the equilibrium curve remains
the same, the number of ideal contact stages can
be varied by changing the slope of the operating
line (L/G).  Remember that when either L or G are
varied, this not only changes the slope of the
operating line, but also changes the locations of
points $x_1$, $y_1$ and $x_2$, $y_2$ because of resulting
concentration changes.  In a counter-current
absorber decreasing L:  decreases the slope;
increases the concentration of $x_1$; increases the

concentration of $y_2$; and it may or may not have any effect on contact time (decreasing G will obviously increase the contact time).

Actual absorber average stage efficiencies can be established by dividing the number of contact stages built into the absorber by the number of theoretical contact stages estimated by the stepping off procedure. When the absorber has no plates (e.g., packed towers), actual efficiencies can be estimated by the procedures given in Section 11.5.4.

## 10.3.7 MASS TRANSFER COEFFICIENTS AND EFFICIENCY

Absorption systems can be either gas phase or liquid phase controlling depending on which film offers more resistance to the molecules being absorbed. Gas phase controlling means that the mass transfer through the gas film is slow compared to the liquid film transfer. Gas phase controlling is desirable in that it usually signifies a faster overall transfer rate than when liquid phase is controlling because the molecules usually pass through the gas film faster.

Calvert (2) reports the following rules: gas phase controls when the value of $k_G$ approaches the value of $K_G$ (the values of $k_G$ are larger than values of $K_G$). Gas phase controls when $k_G$ ≲ 1.1 $K_G$. Low values of $k_G$ means there is poor transfer through the gas phase--this occurs when values of H   3 atm ft$^3$/(lb mole). Liquid phase controls when $k_G$ > 10 $K_G$. Large values of $k_G$ mean good transfer through the gas film; this usually occurs when H > 3,000.

The various mass transfer coefficients for absorption can be calculated using diffusivities. The individual gas film coefficient in lb moles/ (ft$^2$ atm hr) is calculated by:

$$k_G = \frac{14,800}{RT} \left( \frac{D_{AB}}{\pi\, t_G} \right)^{1/2} \tag{10.32}$$

and the individual liquid film coefficient in units of lb mole/(ft$^2$ hr lb mole/ft$^3$) is estimated by:

$$k_L = 236 \left( \frac{D_{AB}}{\pi\, t_L} \right)^{1/2} \tag{10.33}$$

where:   t = contact time, sec

R = gas constant, $\dfrac{82.05 \text{ atm cm}^3}{\text{g mole } °\text{K}}$

T = °K

$D_{AB}$ = diffusivity (properly chosen for gas or liquid phase), cm²/sec

Diffusivities can be estimated from Equations 10.1 and 10.2. If necessary, the conversion factor of molar density of liquid which equals 1b moles total per ft³ should be used. The value of the gas diffusivity will be approximately 0.1 cm²/sec for low concentrations in air and the value of the liquid diffusivity should be approximately 1.5 x 10⁻⁵ cm /sec for low concentrations in water. The contact time ranges from 0.01 to 0.5 seconds. The values of the overall transfer coefficients can be obtained when Henry's Law applies using Equations 10.13 and 10.14 which are rewritten here:

$$\frac{1}{K_G} = \frac{1}{k_G} + \frac{H}{k_L} \qquad (10.13)$$

$$\frac{1}{K_L} = \frac{1}{k_L} + \frac{1}{Hk_G} \qquad (10.14)$$

Predicted absorber contact stage efficiencies for gas phase controlling systems can be estimated using Figure 10.12 which plots contacting efficiency against a term called "overall gas transfer units" for both co-current and counter-current systems. Numerical values of the overall gas transfer units can be estimated by dividing the numerical value of $y_1$-$y_2$ by the y magnitude of the average driving force. The magnitude of this driving force is obtained by subtracting the y value at the midpoint of the operating line from the y value obtained when the driving force line intersects the equilibrium curve--see Figure 10.5. It will require fewer contact stages to separate a system with a small number of overall gas transfer units than for one having a greater number of transfer units. The contact stage efficiencies shown in Figure 10.12 are optimum values which can only be obtained by proper design and operation of an absorber.

Addition of complexing agents to the absorbing liquid to tie up the absorbed pollutants by chemical reaction increases the driving force

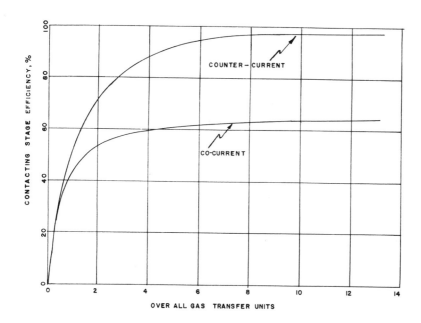

Figure 10.12   Contact Stage Efficiencies for Gas
Phase Controlling Absorption (for
optimum operation)

available which increases the absorption rate. This
is like reducing the back pressure due to concentration and results in a lower number of transfer units
required. This is chemical absorption (chemisorption),
and increases the value of $k_L$ by up to 12 times the
value experienced when only physical absorption
exists.

10.4  GAS ADSORPTION

Adsorption is the taking up of a gas (or liquid
or dissolved substance) on the surface of a solid
or a liquid. (Liquid surface adsorption is not
greatly significant and is not discussed further
in this section.) The surface of any solid contains
some adsorbed material. Certain finely divided
solids such as activated carbon and silica gel
adsorb large quantities of materials because of the
great amounts of surface area available and because
of the surface properties. Both physical attraction

and chemical reaction can take place during adsorption
giving rise to the terms "chemical" and "physical"
adsorption.  Physical adsorption consists of
attracting gas molecules, usually by electrostatic
forces which result from gas molecule polarity and
strongly positive or negative ions on the surface
of the adsorbing solid.  Chemical adsorption
usually consists of physical adsorption accompanied
by a chemical reaction.  Physical adsorption is
most important for gas separation work.

In adsorption, the term *adsorbent* means the
adsorbing solid and *adsorbate* is the adsorbed
material.  Adsorption processes may consist of
contacting the solid with a gas mixture or vice
versa to remove any or all of:  odor, taste,
moisture, solvents or other pollutants from the gas.
The adsorbed species (and the adsorbent) may or
may not be recovered by regeneration of the adsorbent.  Molecular sieves have greatly increased the
use of adsorption processes because these sieves
can be tailor-fitted to adsorb particular size and
type materials (molecular sieves are alkali-metal
silicates).  An example of how molecular sieves are
tailor-fitted for specific applications can be seen
in Table 10.2 which lists some Union Carbide
Corporation molecular sieves (made by Linde
Division).

TABLE 10.2   MOLECULAR SIEVES (LINDE TYPE)

| Type | Nominal Pore Diameter, $\mu$ | Molecules Adsorbed** | Remarks |
|------|------------------------------|----------------------|---------|
| 3A | 0.003 | <3 A° effective diameter (e.g. $H_2O$ & NH ) | used for drying and dehydration. |
| 4A | 0.004 | <4 A° diameter (e.g. ethanol $H_2S$, $CO_2$, $SO_2$, $C_2H_4$ & $C_2H_6$) | scavenge water from solvents & sat. hydrocarbons (HC). |
| 5A | 0.005 | <5 A° diameter (e.g. $n$-$C_4H_9OH$, $n$-$C_4H_{10}$, $C_3H_8$ to $C_{22}H_{46}$) | separate n-paraffin from branched & cyclic HC. |
| 10X | 0.008 | <8 A° diameter (e.g. iso-paraffin & olefins, $C_6H_6$) | separates aeromatic HC. |
| 13X | 0.010 | <10 A° diameter | drying, $H_2S$ & mercaptan removal (gas sweetening). |

**Each type adsorbs listed molecules
plus those of preceding types

The advantage of gas adsorption lies in the fact that frequently it is economically possible to purify gases containing only small amounts of pollutants using adsorbers. The captured gases can be recovered and/or the adsorbent can be reused if one or more of the following recovery cycles are utilized: 1) temperature gradients, 2) pressure gradients and 3) concentration gradients. These cycles or "swings" require semi-batch type operating systems. The polluted gas stream is introduced into the adsorption system under one or more of the favorable conditions of low temperature, high pressure or high concentration. The purified gases leaving the adsorber can be released into the atmosphere until the adsorbent becomes nearly saturated. When the polluted gas stream "breaks through", it is shut off and a clean air or other gas stream is introduced to strip off the adsorbed pollutants at one or more of the following conditions of high temperature, low pressure or low concentration. The recovery effluent gas stream usually contains a high concentration of the pollutant gases making it possible to collect and use them by some method such as absorption. The entire process can be carried out in either fixed or movable (fluid) bed systems.

The gaseous molecules, when captured by an adsorbent, give off a heat of adsorption making the process exothermic. The heat released varies depending on the magnitude of the electrostatic force of the physical attraction (which depends on the polarity of both the adsorbent and the adsorbate) as well as on the chemical reactions which may take place. Highly exothermic adsorption processes indicate that chemical adsorption has occurred. In these cases, it may not be possible to desorb the gases for recovery of either the gas or the adsorbent.

In addition to the factors mentioned above which improve adsorption (low temperature, high pressure and high concentration), adsorption is also improved by: good distribution--which is necessary to provide a high adsorbate concentration near the adsorbent surface; increased adsorbent surface area to provide sites for the gas to be deposited upon; proper selection of adsorbent for removal of desired components and exclusion of components not necessary to adsorb (which would overburden the adsorbing system); pre-treatment to remove high concentrations of competing gaseous materials by

other, more effective processes to further prevent
unnecessary overburdening of the adsorbing system;
increasing the system holdup time to provide
adequate contact time of the adsorbate with the
adsorbent; and continuous replacement of the adsorbent as it becomes unusable for further adsorption.

## 10.4.1  PROPERTIES OF ADSORBENTS

Adsorption, a surface phenomena pertaining to
the taking up of gaseous molecules on the surface,
requires adsorbents that have a large amount of
surface area for a given bulk volume or weight of
material. Surface areas in the range of $2 \times 10^5$
$ft^2/lb$ or more are desired. Some of the surface
area is obtainable by simply breaking up the
material into smaller and smaller sizes. The
surfaces of adsorbents when magnified are highly
irregular which greatly increases the total effective
surface available for adsorption. Adsorbents
typically have sizes ranging from 4 to 20 U.S. Sieve
Series mesh (which is 4.76 mm to 841 micron sieve
opening). Particles smaller than the 20 mesh size
are usually called powders while material from 20
to 4 mesh size are frequently described as pellets
or beads. In addition to the high surface area
required, this material should have a high
oxidation temperature to prevent burning up if
contacted with high temperature gases.

Common types of adsorbents are the carbons,
non-metallic oxides, metallic oxides and combination
oxides. Carbon is one of the older adsorbents and
has enjoyed a high degree of popularity because of
its versatility, economy, and availability. Adsorbent carbon is known by names such as activated
carbon, active carbon and activated charcoal--all
the same thing. It can be made from soft coal,
fruit pits, hardwood and other organic material
of direct vegetative origin, or it can be synthesized
commercially from the reduction of other organic
chemicals.

Activated carbon is produced by heating to
about 900°C in a reducing atmosphere. This produces
the porous particle structure desired with a density
range of from 0.08 to 0.5 grams/$cm^3$. Selective
sizing of this material to obtain the small 70-80
mesh material would result in particles with an
effective mean diameter of about 250 microns.
Assuming a spherical configuration, a quick
calculation would indicate that the surface area

is approximately 900 square feet per pound. The
estimated internal surface area of active carbon
(additional surface area due to the irregularity
of the surface) is in the vicinity of $16 \times 10^5$
$ft^2/lb$. This shows that *total* surface area of an
adsorbent consists of mainly the internal surface
area plus the small amount of normal external
surface.

Many of the non-metallic oxides can be readily
obtained because they exist almost naturally.
For example, diatomaceous earths, which are also
known by the names kieselghur and the trade name
Sil-O-Cel, are obtained from the calcification of
diatomite, which is a soft, earthy rock composed
of siliceous shells of small aquatic plants.
Although the true specific gravity of this material
is approximately 2, the apparent specific gravity
of the finely divided calcined form is less than
0.5. Fuller's earth is another non-metallic oxide.
It is a clay-like powder which can be used after
drying directly in its natural state. Silica gels
are amorphous (non-crystalline) silica which are
produced commercially by the reaction of sodium
silicate and sulfuric acid. These adsorbents
have pore sizes ranging from 20 to 10,000 A°.
Commercial silica gels, like carbon and the other
non-metallic oxides, are not capable of withstanding
high temperatures. Silica gels will break down
at temperatures of 500°F and above.

Metallic oxides have the advantage of being
able to withstand higher temperatures. However,
these materials are not used alone as adsorbents
because they are not electrophilic enough (i.e.,
they do not "like" electrons). They are insulators
and, therefore, must be modified by the addition
of ionic binders. Activated alumina is a highly
porous and granular form of aluminum oxide with a
sp.gr. of 0.8. It is popularly used for the adsorption of moisture from gases. This material can be
reactivated by heating to 350-600°F to drive off
the moisture.

The combination oxides consist almost entirely
of molecular sieves which are crystalline zeolites
containing metallic cations such as sodium. Zeolites are a class of hydrated silica and/or alumina
molecules that occur naturally or can be produced
artificially. Molecular sieves can be obtained or
produced that have various nominal pore diameters
ranging from 3 to 10 Angstroms (see Table 10.2).
Gas molecules with effective diameters smaller

than the sieve pore openings can enter the pores
and become trapped there.   Table 10.3 summarizes
some physical properties of the various adsorbents.
Pore volume is determined by measuring the volume
of helium gas required to fill the voids *and* pores
of an evacuated adsorbent and subtracting from that
the volume of mercury required to fill only the
voids.

### 10.4.2   MODIFIED ADSORBENTS

The addition of other chemicals to an adsorbent
can be used to promote a chemical reaction that
can tie up the adsorbate with the additive.   This
induced chemical adsorption can greatly increase
the rate of adsorption as well as increase the
capacity of the system, but if not properly chosen,
can lead to degradation of the adsorbent causing
it to be structurally weaker and to wear away
quicker due to attrition from the gas flow.   The
additive can also react chemically with the
adsorbent, especially at higher temperatures, which
could weaken and ultimately destroy the effectiveness
of the adsorbent.
Occasionally, it is possible to impregnate an
adsorbent with a catalyst which can increase the
rate of chemical reaction between the various
pollutants that are adsorbed.   Catalytically
stimulated oxidation reactions with air are
examples.   It must be remembered that the modified
adsorbents are often not capable of being regenerated.
For example, lead acetate impregnated on carbon
causes chemical adsorption of hydrogen sulfide
from a gas stream.   The resulting lead sulfide
deposit on the carbon cannot be recovered and the
carbon adsorbent must be discarded when saturated.

### 10.5   OTHER CHEMICAL REMOVAL PROCESSES

The use of chemical reactions in conjunction
with the removal of gaseous pollutants has been
considered in this chapter under both gas adsorption
and absorption.   In addition, chemical reactions
occur during combustion-related disposal operations
as discussed in Chapters VI (Automotive Pollution)
and VIII (Combustion Related Disposal).   There are
also the photochemical types of chemical reactions
included in Chapter IV.   In addition to all these,
chemicals are added directly to gas streams for
pollution control.

TABLE 10.3  TYPICAL PROPERTIES OF ADSORBENTS

| Adsorbent | Form | External Surface Area, $ft^2/lb$* | Pore Volume $ft^3/lb$ | Reactivation Temp., °F | Max. Gas Flow, $cfh/lb$ | Sp. Heat, $C_p$ $Btu/(lb°R)$ | Typical Adsorbates |
|---|---|---|---|---|---|---|---|
| Activated carbon | pellets | 10.5-21.5 | 0.010-.013 | 200-1000 | / | 0.25 | $CH_4$ through |
| | beads (G) | 15.0-24.0 | | 200-1000 | / | 0.25 | $nC_5H_{12}$, $CO_2$, $H_2S$. |
| Silica gel | beads (G) | 5.0-16.0 | 0.007 | 250-450 | 75 | 0.22 | $CH_4$ through $C_4H_{10}$, $C_2H_4$ |
| | beads (S) | 6.0 | | 300-450 | 75 | 0.25 | through $C_4H_8$, $H_2O$, $SO_2$, $H_2S$ |
| Activated alumina | beads (G) | 7.0-18.5 | 0.006 | 350-600 | 50 | 0.22 | $H_2O$, $H_2S$, |
| | beads (S) | 4.0-8.0 | | 350-1000 | 50 | 0.25 | oil vapors. |
| Molecular sieves | pellets | 9.0-14.5 | / | 300-600 | / | 0.23 | See Table 10.2 |
| | beads (G) | 32.0 | / | 300-600 | / | 0.23 | |
| | beads (S) | 7.5-12.5 | / | 300-600 | / | 0.23 | |

G = Granules
S = Spheroids
* Not including the internal pore surface area which is most of the surface (see Section 10.4.1)

## 10.5.1   DIRECT REACTIONS

Gaseous pollutants can be removed from gas streams by the direct injection of reactive chemicals. In these cases, the reaction products are usually precipitates which can then be removed by one of the particulate removal methods. The chemicals used can be either liquids, solids or gases. An example of a solid chemical additive is calcium oxide (calcined limestone) which is injected into gases containing $SO_2$ to produce calcium sulfate, a precipitate. Ammonia is an example of a gas that can react with $SO_2$ and moisture in gas streams to produce ammonium sulfite or ammonium sulfate which are also precipitates.

## 10.5.2   ION EXCHANGE

Ion exchange chemical reactions are significant in water pollution control but have no real application in air pollution control because atmospheric aerosols have varying charges (they may be either positive, negative or neutral). Ion exchange in air pollution work is mainly limited to analytical procedures.

## 10.6   OTHER PHYSICAL REMOVAL PROCESSES

Physical methods are significant for gaseous and particulate pollution control. In this chapter, we have already discussed the significance of physical processes in both absorption and adsorption. In addition to these, it is possible to "purify" gases by diluting them or by masking the smell of odorous pollutants using chemical additives.

Odors are the most frequent source of air pollution complaints. Odors must be evaluated by the human nose, which is an inexact instrument. The sensitivity of the nose varies from day to day and there is wide variance between individuals exposed to the same odor stimulus. Unfortunately, highly odorous materials are observed at very low concentrations in the atmosphere. We have already mentioned that ammonia, which appears to be highly odorous, can be detected at concentrations of approximately 47 ppm while isovaleric acid can be detected by some people at concentrations as low as 0.6 ppb (3). In addition, ethyl mercaptan and butyric acid have odor thresholds of 1 ppb and

hydrogen sulfide has a threshold of only 0.5 ppb (4).
At these extremely low concentrations, it is
*sometimes* more practical and just as effective to
use a physical method of masking or diluting the
smell rather than to actually remove the pollutant
from the air.

## 10.6.1 DILUTION

Some air pollution control regulations state
that the maximum concentration of an odor should
not exceed an indicated number of dilutions. The
number of dilutions permitted, for example three,
means that when the odorous air is diluted with
three times its volume of odor-free air the odor
should no longer be perceptible. There is actually
little significant difference between dilution
values from two to five because the nose is usually
incapable of distinguishing differences in these
regions. The sensitivity of the nose increases
exponentially when we consider odor sensitivity
versus dilution volumes.

A commercially available device which utilizes
the nose as the sensory receptor is made to operate
on the principle of dilutions. This device, called
a Sentometer, is devised so that two independent
air streams can be judged by the nose. One stream
is the ambient air and the other is a stream of
air that has been purified by passing through
a charcoal adsorber. The ambient air can be diluted
by measured volumes of purified air from the
adsorber until no more odor is detectable.

## 10.6.2 MASKING

Odor masking implies that the quality of the
original malodor is obscured while little dilution
of odor concentration actually occurs. Masking
is one of the only methods that can effectively be
used with both stack type emissions and unconfined
large source odor systems. When used with stack
type emissions, masking is frequently used in
addition to removal preocesses which do not completely
remove the odor. Unconfined odors are extremely
difficult to treat. Masking is sometimes a
convenient method of helping to solve an undesirable
odor problem that originates from large unconfined
sources such as waste treatment plants, open
ditches, lagoons and chemical operations where it
is not normally practicable to collect and treat
the gases by passing them through a gas purification
system.

Masking chemicals are typically perfume type substances. Caution should be exercised when masking odors to be sure that 1) no poisonous substance is masked and 2) the masking chemical itself does not become offensive and create an odor problem. Although odors are caused by gases or vapors, it should be pointed out that particulate matter can produce and/or carry odors. It is necessary to remove the particulate matter to remove the odors.

### 10.6.3   GOOD HOUSEKEEPING

Before leaving the subject of physical removal processes, it can be pointed out that the simple factor of good housekeeping is *absolutely* necessary before pollution can be controlled. This obvious, yet too frequently neglected physical process consists of cleaning up, disinfecting, confining, etc. In actual practice, watch for this and be sure it is carried out!

### 10.6.4   DISTILLATION AND FREEZE CONCENTRATION

The chemical operations of distillation and freeze concentration are seldom used for air pollution control per se. The reason for this is that the concentration of pollutants in the waste gas streams are usually too dilute for these processes to be economically practical. More importantly, these operations, when not properly performed, contribute to air pollution and should be kept under the most effective controls to help reduce losses into the atmosphere.

### QUESTIONS FOR DISCUSSION

1. What is the difference between chemical and physical separation processes for gas purification?
2. Physically interpret the meaning of "diffusivity" in the units of $distance^2/time$.
3. The expressions $N_A$ and $\Delta N_A$ are used in this chapter. Is there any difference in meaning between them?
4. How should one initially attempt to determine which gaseous purification method should be utilized for a given pollutant?
5. Describe an experiment to make surface force measurements.
6. Describe an experiment to obtain vapor-liquid equilibrium data.

7. Make a mass balance on a co-current absorption system to obtain the equation for the operating line. State all assumptions.
8. How does an adsorber work?
9. What size molecular sieve must be used in order to remove benzene? What other components would also be removed?
10. What approximate percent of the total surface area of activated carbon is the internal surface?

PROBLEMS

10.1 Calculate the diffusivity of $SO_2$ in air at standard conditions.
10.2 Calculate the diffusivity of $SO_2$ in water at 70 F.
10.3 At what temperature will a water solution containing 15% glycerine boil under normal conditions?
10.4 Estimate the value of $k_L$ and $K_L$ for the absorption of CO from air by water when the temperature is 80°F and the pressure is 800 ppm. When there is 0.1 cc CO disolved per 100 ml water, $10^{-3}$ lb moles CO/hr can be absorbed by a 5 ft² area and $k_G$ is 0.16 moles/(hr ft² atm). Is this gas phase or a liquid phase controlling absorption?
10.5 An industrial chemical operation makes aqueous ammonia by absorbing with water. The absorber effluent gas stream contains 0.025 mole $NH_3$ per mole of gas which results in excessive odor in the ambient air. A second absorber is to be designed which will utilize the effluent from the first absorber as the input. The exit gas stream from this second absorber should contain ammonia at a maximum concentration of 500 ppm. Use the following conditions: water is the absorbent, the inlet gas flow at SC is 1600 cfm, slope of the operating line is 1.5 (therefore the system is counter-current). If ammonia vapor-liquid data are not available, use the following equation for the equilibrium line:

$$y = 0.77x \qquad (10.34)$$

How many theoretical stages of separation are required for this absorber and how much water is required?
10.6 The maximum allowable concentration for $SO_2$ in the ambient air is 0.015 ppm. An exit gas stream contains 5% $SO_2$. It is desired to construct a 20 foot high absorbing tower to reduce the $SO_2$ in the exhaust gas to an acceptable level. Assume there will be one plate per foot of column height in the

absorber, absorption efficiency is 100% (i.e., actual
plates equal theoretical plates) and that the
absorber will discharge into the atmosphere at a
height of 30 meters. It may also be assumed that
the equilibrium of $SO_2$ in water at 20°C follows
the curve:

$$y = 33x \qquad (10.35)$$

Estimate the minimum amount of water required to meet
these requirements if the gas flow is 600 cfm at SC
and flows are counter-current.

10.7  If the plates are removed in the absorber of
Problem 10.6 and the column is filled with packing
(assume the same absorption efficiency and no. of
theoretical plates), what surface area is required
for the absorption? (Hint:  Calculate $K_G$ using
Henry's Law data and procedure in Section 10.3.7)

10.8  Estimate the value of $K_G a$ for the data given
in Problem 10.6. How does this compare with the
values obtained in Problem 10.7? The gas velocity
in 10.6 is 1 ft/sec.

10.9  Pore capacity of adsorbents can be measured
by passing a measured amount of helium into a
previously evacuated amount of adsorbent, then
filling the system with a measured amount of mercury at atmospheric pressure. (The mercury does
not enter the pores.) If the voids between the particle
particles equal 1/3 of the total volume, how
much helium and mercury would be required for this
test using 2 pounds of activated carbon with a
specific gravity of 0.4? Assume a pore volume of
$0.012 \text{ ft}^3/\text{lb}$.

10.10  A 1000 cfm dry waste gas stream contains
0.05% methane. Assuming that the breakthrough
occurs when the adsorbent is 65% saturated, what
volume and weight of activated carbon is required
for each bed if the bed is to be kept onstream for
a maximum of 1/2 hour?

10.11  Repeat Problem 10.10 using silica gel if
the gas stream also contains 0.03 lb $H_2O$ vapor per
lb dry air.

REFERENCES

1.  Gilliland, E.R., Industrial and Engineering
    Chemistry, Vol. 26, p. 68 (1934)
2.  Calvert, S., "Source Control by Liquid Scrubbing,"
    Air Pollution, Ch. 46, Ed. by A.C. Stern,
    Academic Press (1968)

3.  Hanna, G.F., "Odor Measurement Methods," AIChE Odor Control Symposium, Atlanta, Ga., February (1970)
4.  Leonardos, G, et. al., "Odor Threshold Determinations of 53 Odorant Chemicals," JAPCA, Vol. 19, No. 2, pp. 91-95 (1969)

# CHAPTER XI

# CONTROL EQUIPMENT

This chapter includes the design of various types of equipment used in air pollution control, their efficiency and application. Some of these are used to control both gaseous and particulate emissions, and sometimes several different physical and/or chemical operations take place in a single type device. Because control theories for particulates and gases were discussed independently, control equipment now can be considered according to its type. In this way, advantages as well as disadvantages can be more easily compared.

The first section deals with a method of determining collection efficiency for particulate matter which can be applied to any type particulate control device. Efficiency for gas recovery has a less generalized procedure and each case must be treated independently for the specific gas, liquid or solid, and collector operating conditions utilized. Keep in mind as you progress through this chapter that efficiency is also a function of operating technique and power input to the collector, in addition to the factors being presented. Section 11.8 presents some control variables related to collection devices already considered in previous chapters.

## 11.1 OVERALL EFFICIENCY

It is possible to determine overall efficiency for any pollution control device by dividing the amount of pollutants captured by the amount of pollutants available to be captured. Where it is expressed as a fraction or percent overall efficiency ($E_0$), it must be expressed in consistent terms (*e.g.*, by mass or volume, area or soiling

index, number, etc.).  It is desirable to state
the units from which the efficiency was determined.
In the absence of this statement, assume that
collection efficiency of particulate matter is
expressed by weight, collection efficiency of gases
expressed by volume.
   It is desirable, from the equipment manufacturer's
point of view, to express efficiencies of particulate
collectors as percent by weight because this makes
the particular device appear to have a higher
efficiency.  The same device will show successively
lower efficiency values when reported by area and
by number (in that order).  There are no convenient
standard curves for collection efficiency of gases,
though collection efficiency curves for particulate
collectors can be obtained from the manufacturer or
by experimental data.  Efficiency can also be
approximated using typical curves (such as those
presented at the end of this chapter in Section
11.9) which are typical for the given families of
devices.  These curves will in general be similar
to Figure 11.1a which shows collection efficiency
of some given device for specific particle
diameters.
   Chapter VII shows that particulate matter in
exhaust gases usually has a log-normal size
distribution which can be expressed on probability
type coordinates as a straight line.  It can also
be plotted to give curves such as Figure 11.1b,
which is a plot of cumulative weight percent
undersize versus particle diameter for some
unspecified dust.  Comparing Figure 11.1a with
Figure 11.1b makes it apparent that overall
collection efficiency of particulate matter is a
function of both the type of collection device and
the type (size and distribution) of dust.  This
can be expressed mathematically as:

$$E_o = \int_o^\infty f(E)\, d(d)$$

(11.1)

where:  $E_o$ = the determined efficiency for a
          specific diameter
        $d$ = differential operator and (d) is
          particle diameter

   Instead of trying to obtain analytical
expressions for the terms in Equation 11.1, it is
usually more convenient to obtain overall
efficiency by the graphical procedure.  This consists

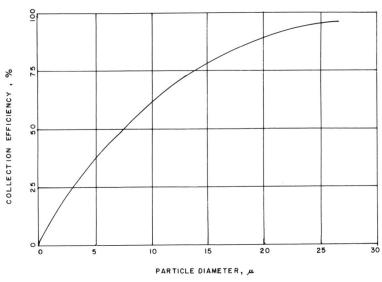

(a) A TYPICAL COLLECTION EFFICIENCY vs PARTICLE DIAMETER CURVE

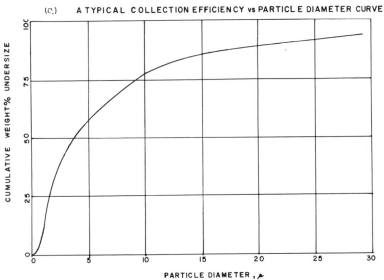

(b) PATICLE SIZE DISTRIBUTION FOR SOME "UNSPECIFIED" DUST

Figure 11.1  Particle Efficiency and Distribution Curves Required to Determine Overall Efficiency

of plotting cumulative weight percent undersize versus collection efficiency for various particle diameters. Figure 11.2 shows this by combining the data of Figures 11.1a and b. The overall efficiency for the system then can be estimated by balancing the area under the curve on the left to an equal area above the curve on the right, as in Figure 11.2. For this example, the estimated overall efficiency ($E_O$) for this collector and dust is approximately 38%. This method of estimating overall efficiency is usually accurate enough for design purposes. In practice, it is best to take measurements and calculate the overall efficiency directly for the given system and operating conditions.

Sometimes it is desirable to place collectors in series to obtain more complete removal of pollutants. Overall efficiency for the system can be calculated using physical measurements if the system is operating; it is also possible to estimate overall efficiency using the overall efficiencies

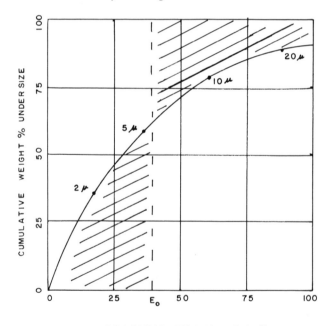

Figure 11.2    Graphical Estimation of Overall
                Particulate Collection Efficiency

of each individual device. In this case, the equation for a series of collectors becomes:

$$E_0 = E_{0,1} + (1-E_{0,1})E_{0,2}$$

$$+ (1-E_{0,1})(1-E_{0,2})E_{0,3} + \cdots$$

(11.2)

where: $E_{0,1}$ = overall collection efficiency of first collector
$E_{0,2}$ = overall collection efficiency of second collector
etc....

Equation 11.2, when written for n units, becomes:

$$E_0 = [E_{0,1}] + [(1-E_{0,1})E_{0,2}]$$

$$+ \cdots + [(1-E_{0,1})(1-E_{0,2}) \cdots$$

$$(1-E_{0,n-1})E_{0,n}]$$

(11.3)

Collectors in parallel are used to obtain greater thru-put rates. Overall efficiency for these systems can be estimated using weighted individual unit overall efficiencies.

## 11.2  SETTLING CHAMBERS

Settling chambers are simple devices used for particulate collection only. These devices, which may have various configurations, utilize gravity to separate particulate matter from an effluent gas stream. They are constructed with a large cross-sectional area perpendicular to gas flow so that the gas velocity is reduced to a minimum in an attempt to prevent re-entrainment of the settled particles.

The two basic types of settling chambers are shown in Figure 11.3. The simple chamber is a single compartment (large box) with a sloping bottom to direct the collected dirt to a common discharge point. The Howard Separator is a baffled chamber with horizontal plates spaced one inch (minimum) or more apart. As would be expected, this separator is more efficient although it is also more costly to build and not as easily cleaned. (It is cleaned by blowing down or washing off the plates.)

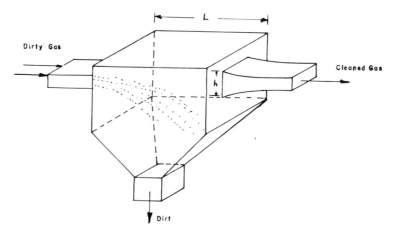

(A)   SETTLING   CHAMBER

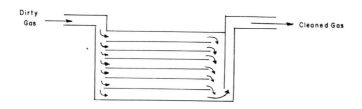

(B) HOWARD SEPARATOR

Figure 11.3   Basic Settling Chambers

The dirt which enters the chamber with the inlet gas stream initially has a velocity equal to the velocity of the gas in the inlet duct. When the gas velocity is reduced to essentially "zero" by the expansion into the chamber, the particles have two velocity components, $v_x$ in the horizontal direction and $v_y$ in the vertical direction as shown in Figure 11.4.  The forces acting in the horizontal direction are the kinetic energy and drag forces which cause the particle to slow down and stop according to Stokes' stopping equation (see Equation 9.11 and Figure 9.4)

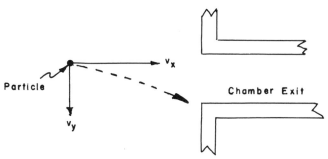

Figure 11.4    Particle Trajectory in Settling
Chamber

Ideally, the force in the vertical direction is
due solely to the force of gravity, but air
movement modifies this (see Section 9.1).  Particle
trajectory is indicated by the dashed line and
shows that if the particle does not fall below the
lower edge of the chamber exit, it will be swept
out and not collected.
    As just mentioned, the ideal design would be
to have the horizontal velocity of the gas in the
chamber equal to zero.  However, this is not
practical and, in fact, velocities low enough to
provide viscous flow are not usually practical.
The maximum velocity recommended is about 10 ft/sec,
with 1 ft/sec being a commonly used velocity.
Because of this velocity, there are turbulent
disturbances in the chamber which effectively slow
down the rate of settling so the value of one half
the Stokes' terminal free fall velocity is assigned
to $v_y$.  Using these assumptions, the following
general formula can be derived which gives the
diameter of the smallest particle that can be
*completely* removed in a gravity separator:

$$d_m = \left[ \frac{36 \ v_x{}' \ h'\mu}{\rho \ g \ L} \right]^{\frac{1}{2}}$$

(11.4)

Equation 11.4 can be rewritten for English units of
distance and velocity as:

$$d_m = 3.5 \times 10^5 \left[ \frac{v_x h \mu}{\rho \ g \ L} \right]^{\frac{1}{2}}$$

(11.4a)

where:  $d_m$ = minimum diameter particle
             collected, microns
        $v_x$ = initial particle horizontal
             velocity, ft/sec
        h = height of outlet, feet
        L = horizontal length of chamber, ft
        $\mu$ = gas viscosity, g/(cm sec)
        $\rho$ = particle density, g/cm$^3$ ·
        g = gravitational acceleration $\simeq$
             980 cm/sec$^2$

The efficiency of a settling chamber for
particles of various specific diameters can be
estimated using:

$$E = \frac{0.5 \; v_s' \; L}{v_x h} \qquad\qquad (11.5)$$

where:  E = specific efficiency (fraction) for
             particle size used to obtain
             Stokes' terminal falling velocity
             ($v_s'$).  Units of $v_s'$ are ft/sec.

(Be certain to use consistent units throughout for
Equation 11.5.)  Note how the Howard chamber has a
much greater efficiency because of the greatly
increased ratio of L/h.

Settling chambers are frequently used to provide
low efficiency but economical pre-cleaning of
exhaust gases on power plant boiler exhaust, fume
manufacturing processes and spray chamber waste
gases.  It is possible to spray water into these
devices to achieve increased particle collection
as well as to obtain some gaseous collection.

## 11.3  CENTRIFUGAL SEPARATORS

Simple centrifugal separators are used for
particulate collection.  However, like the settling
chambers, they can be used to remove gaseous
pollutants only if they are modified by the addition
of sprays or some other liquid scrubbing facilities.
Cyclones make up the largest majority of centrifugal
separators, though blowers and other devices that
utilize centrifugal action can be classed in this
section.

Cyclone separators have gas entrances designed
to cause the gas to swirl inside the cylindrical
body.  The gas and particles both have essentially

the same tangential velocity, but the particles
have a much greater normal velocity due to
centrifugal force and high particle density. This
forces the particles to move outward toward the
cyclone wall, where those that reach the wall are
separated from the gas stream when they fall to
the bottom of the cyclone.

It is estimated that a gas stream makes 5 to
10 revolutions in the cyclone and then passes up
through the center of the separator where it
leaves through a top outlet. The inner gas vortex
diameter is estimated to be less than twice
the diameter of the gas outlet. The bottom of a
cyclone separator is usually conical so that the
collected dirt and/or liquid can be conveniently
removed at a common bottom outlet. Standard
cyclone design dimensions are given in Table 11.1
and refer to the dimensions indicated in Figure 11.5.

Design variations are commonly made to achieve
specific results. For example, rotary vanes are
sometimes added to prevent updraft when collecting
low density particulate matter. Cyclones are
irrigated by liquid sprays to facilitate removal
of sticky solids. (Otherwise, they are pounded
on·by manual or mechanical impulse systems.) Inlet
vanes are frequently added in attempts to increase
the centrifugal force. Pressure drop in a cyclone
can be estimated by the following equation, which
takes into account the presence of vanes and the
positioning of these vanes.

TABLE 11.1   STANDARD BASIC CYCLONE DESIGN
             DIMENSIONS

| | *Conventional* | *TYPE* High *Thruput* | High *Efficiency* |
|---|---|---|---|
| Cylinder diameter, D | D | D | D |
| Cylinder length, $L_{cy}$ | 2D | 1.5D | 1.5D |
| Cone length, $L_{co}$ | 2D | 2.5D | 2.5D |
| Total height, H | 4D | 4D | 4D |
| Outlet length, $L_o$ | 0.675D | 0.875D | 0.5D |
| Inlet height, $L_i$ | 0.5D | 0.75D | 0.5D |
| Inlet width, $W_i$ | 0.25D | 0.375D | 0.2D |
| Outlet diameter, $D_o$ | 0.5D | 0.75D | 0.5D |

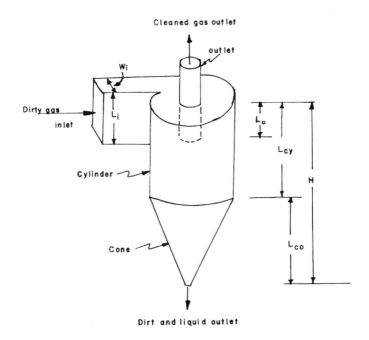

Figure 11.5   Typical Cyclone Separator

$$\Delta P = \left[ \frac{0.0027QF}{C \, D_o^2 \, W_i L_i \left(\frac{L_{cy}}{D}\right)^{1/3} \left(\frac{L_{co}}{D}\right)^{1/3}} \right] \left[ \frac{1}{0.013C_i + 1} \right]$$

(11.6)

where:   $\Delta P$ = pressure drop, inches $H_2O$
         $Q$ = volumetric flow rate, cfm
         $C_i$ = inlet dust load, grains/ft$^3$
         $C$ = 0.5 for no vanes
           = 1.0 for type "a" vanes (Fig. 11.6)
           = 2.0 for type "b" vanes (Fig. 11.6)
         $D, D_o,$
         $L_{cy}, L_{co},$
         $L_i, W_i$ = cyclone dimensions, ft
         $F$ = 0.8 for high thruput cyclones
           = 1.0 for conventional cyclones
           = 2.2 for high efficiency cyclones

This equation is modified by the Briggs' method (1)
of accounting for dust loading.  Note that $\Delta P$
decreases as dust loading increases.  This equation
does not show that $\Delta P$ can be reduced if the cyclone
can be constructed using smooth materials.

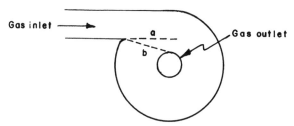

Figure 11.6   Cut-away Top View of Cyclone with Vanes

Cyclone separators are economical pre-cleaners.
Their efficiency is low, but they are inexpensive,
economical to operate devices which also have the
advantage of being able to handle high temperature
gases.  They can be used singularly, in series and/
or in parallel.  Series operation results in
greater cleaning of the waste gases while parallel
operation provides increased system capacity.
Typical operating conditions for cyclones are:

| | |
|---|---|
| Inlet gas velocity | - 50 ft/sec (20-70 ft/sec range) |
| Gas volumetric flow rate | -30 cfm to 50,000 cfm per unit |
| | - 375 $D^2$ ft$^3$/min (conventional) |
| | - 840 $D^2$ ft$^3$/min (high thruput) |
| | - 300 $D^2$ ft$^3$/min (high efficiency) |

where:  D = cyclone diameter in feet

| | |
|---|---|
| Gas temperature | - 750$^O$F maximum |
| Pressure loss* | - 0.5 to 2.0 incles of water (conventional) |
| | - 2 to 6 inches of water (high efficiency and multiple) |

*Proportional to flow$^2$

The diameter of the smallest particle ($d_m$) that
can be removed from the inlet gas stream with 100%
efficiency in a cyclone separator can be estimated
by:

$$d_m = 17.3 \times 10^5 \left[\left(\frac{9\mu_g \, D^2 g}{32(\rho_p - \rho_g)v_i \, H}\right)\left(\frac{D^4 - D_O{}^4}{D^4}\right)\right]^{\frac{1}{2}} \qquad (11.7)$$

where:   $d_m$ = microns

$\mu_g$ = gas viscosity, lb sec/ft$^2$

$\rho_g$ = gas density, lb/ft$^3$

$\rho_p$ = particle density, lb/ft$^3$

$v_i$ = inlet gas velocity, ft/sec

g = gravitational acceleration $\simeq$ 32.174 ft/sec$^2$

H, D, and $D_o$ = cyclone dimensions, ft

The diameter of particles that can be collected with a 50% efficiency ($d_{cut}$ in microns) can be estimated by:

$$d_{cut} = 17.3 \times 10^5 \sqrt{\frac{9\mu_g W_i g}{2\pi N v_i (\rho_p - \rho_g)}} \qquad (11.8)$$

where:   N = no. of revolutions of gas stream = 5 to 10
$W_i$ = cyclone inlet width, ft

And the diameter of the smallest particle that can be collected from a given streamline at a specific distance from the cyclone wall ($d_{xm}$ in microns) can be estimated by:

$$d_{xm} = 17.3 \times 10^5 \sqrt{\frac{9\mu_g x g}{\pi v_i H (\rho_p - \rho_g)}} \qquad (11.9)$$

where:   x = distance streamline is from cyclone wall, ft.

Combining equations 11.7 and 11.9 relates the minimum sized particle that can be collected from a given stream line to the smallest particle removed with 100% efficiency (for x greater than 6 inches use the constant x = 0.5):

$$d_{xm} = d_m \sqrt{\frac{2x}{D - D_o}} \qquad (11.10)$$

It is possible to obtain an efficiency equation
from these relationships, but it would be necessary
to express cyclone collection efficiency as a
function of both particle size and distance from
the wall. For this reason, and because of the
effects due to eddy currents, particulate bouncing,
coagulation, re-entrainment, dust loading variations
and particle configuration, it is preferred to use
actual data for cyclone efficiency. Typical
efficiency curves for various cyclone separators
are given in Figure 11.15 at the end of this
chapter.

Abrasion is frequently a problem in cyclone
separators, and the life of some installations can
be improved by the use of low coefficient of
friction materials in construction. For example,
polyvinyl chloride coated steel (PVS) reduces
abrasion as well as reduces the pressure drop in
cyclones handling abrasive solids.

## 11.4  INERTIAL SEPARATORS

This is the largest class of equipment, used
for both particulate and gaseous pollution control;
however, it is only the particles that are removed
by the inertial processes while the gases are
removed by absorption or adsorption. This makes
it difficult to accurately classify these devices
and, for this reason, some inertial separators are
discussed separately. For example, centrifugal
separators are inertial separators and these have
already been discussed. Wet scrubbers are
sometimes inertial separators, and filters are
always inertial separators. Yet it is necessary
to classify these separately and consider them
later in this chapter.

The devices using atomized liquids as the
particle collection targets utilize the liquid to
carry the collected material out of the system.
Material collected on stationary impaction targets
must be removed by mechanical cleaning such as
shaking, impulse rapping, reversing gas flow (blow
down) and periodic shutdown, scraping and flushing.
Packed towers, when used dry, are true inertial
separators, though this is seldom the approach
used in actual practice. More likely, the tower
will be used with a counter-current water flow,
making it not only an inertial separator but an
absorber. Venturi scrubbers are extremely popular
and efficient inertial separators--see the wet
scrubber section, 11.5.7.

Some inertial separators that usually have no
other classification are the louvered collectors
and baffle chambers. Both of these devices utilize
only the inertial force of the particle to accomplish the capture. Figure 11.7a shows that dirty
gas entering the louvered chamber is forced to turn
$90°$ before it can leave. The particulate matter,
with its high inertia, continues moving forward as
predicted by Stokes' stopping equation. Some of
the gases are permitted to leave the bottom of the
louvered collector so that they can carry out the
deposited dirt. The baffle chamber works on
essentially the same principle except that the
entering gas must deflect around a series of
obstacles. Although there are more directional
changes, the gas is not forced to turn as sharply
in the baffle chamber as in the louvered collector.
The baffle chamber is shown in Figure 11.7b.
Collected material can be removed when it falls to
the bottom.

Even though inertial separators are extremely
important air pollution control devices, very
little information is available concerning the
mechanism of particle capture other than the fact
that capture is by inertial deposition. Attempts
to relate what is known about this mechanism can
be made using the inertial impaction parameter ($P_I$).
This parameter is defined by:

$$P_I = \frac{X_s}{R_c} \qquad (11.11)$$

where:  $X_s$ = Stokes' stopping distance =
$\dfrac{2v_i r^2 \rho_p}{9\mu_g}$ , cm

$R_c$ = collector radius, cm
$v_i$ = particle initial velocity, cm/sec
$r$ = radius of particle, cm
$\rho_p$ = particle density, g/cm$^3$
$\mu_g$ = gas viscosity, g/(cm sec)

Therefore, the impaction parameter can be rewritten
as:

$$P_I = \frac{2v_i r^2 \rho_p}{9\mu_g R_c} = \frac{v_i d^2 \rho_p}{18\mu_g R_c} \qquad (11.11a)$$

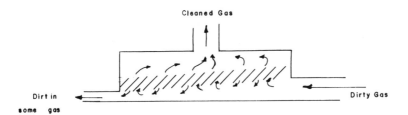

(A) LOUVERED COLLECTOR

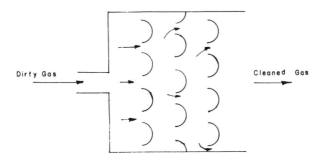

(B) BAFFLE CHAMBER

Figure 11.7   Inertial Separators

For a given collector, it has been pointed out that there is a collection efficiency for each particle size, so for a given collector and operating system:

$$E = f (P_I) \qquad\qquad (11.12)$$

The collection efficiency (fraction) of a dry packed bed is estimated (2) to be related to the impaction parameter by:

$$E = (1 - \exp [ -C \frac{Z}{D} P_I]) \qquad (11.13)$$

where:  $Z$ = height of packed bed, ft
$D$ = diameter of bed, ft
$C$ = 10.7 for 1/2" Beryl or Intalox
saddles, marbles, or Raschig rings
= 10.0 for 1" Beryl saddles or Raschig
or Pall rings
= 12.0 for 1-1/2" Beryl saddles or
Raschig or Pall rings
= 100.0 for 3-5" Coke

Fiber filter collection efficiency for a specific fiber size is approximately equal to the impaction parameter.

$$E \cong P_I = \frac{X_s}{R_c} \qquad (11.14)$$

where:  $R_c$ = fiber radius

This approximation is only true for single layer fibers.

Cyclone separator efficiency can also be related to the impaction parameter. If a particle travels the distance equal to the width of a gas inlet ($W_i$ as shown in Figure 11.5) during the time that it is in the cyclone, theoretical collection efficiency for that particle is 100%. Let the actual distance a particle travels during its stay in the cyclone equal $\delta$. Also assume that the gas drag on the particle moves it through the cyclone at a tangential velocity ($v_t$) approximately equal to the gas inlet velocity ($v_i$). The force of gravity can be neglected, but centrifugal force must be considered as well as the drag forces. Figure 11.8 shows the velocity vectors and forces acting on a particle in a centrifuge. The resultant path of the particle is indicated by the dashed lines. The normal velocity ($v_n$) results from the combination of centrifugal force ($-F_n$) and radial component of the drag force ($F_D$). The tangential velocity is a result of the inertial flow force ($F_F$) and the negative circumferential component of the drag force ($-F_D$). The drag force in both directions is proportional to the difference between the square of the resultant particle velocity and the square of the gas velocity. Under normal conditions, this velocity difference is essentially zero so the drag

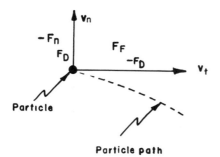

**Particle path**

Figure 11.8   Velocity and Force Vectors on a
Particle in a Cyclone

force can be neglected.  As indicated in Chapter IX,
the separation force in a working cyclone is
approximately 400 which explains why the gravitational force can be neglected.
    The time a particle stays in the cyclone outer
vortex can be estimated as:

$$t = \frac{2\pi r N}{v_i}$$   (11.15)

where:   t = time particle stays in cyclone
             center vortex, sec
         r = particle radius, cm
         N = no. of revolutions of gas stream
         $v_i$ = gas inlet velocity, cm/sec

Noting that an expression for the normal velocity
can be obtained in the form of the Stokes' terminal
centrifugal velocity, we can equate this to the
differential form of the velocity as:

$$v_n = \frac{d\delta}{dt} = \frac{d^2 \rho_p v_i^2}{18\mu R_c}$$   (11.16)

where:   $R_c$ = radius of the streamline

Upon integrating this expression and substituting
for the time, the following expression is obtained
for the distance a particle moves in the normal
direction while in the cyclone:

$$\delta = \frac{d^2 \rho_p v_i^2 t}{18\mu R_c} = \frac{v_i d^2 \rho_p N}{9\mu}$$   (11.17)

The theoretical efficiency can then be expressed as:

$$E = \frac{\delta}{W_i} = \frac{4v_i r^2 \rho_p N}{9\mu W_i} \qquad (11.18)$$

where particle diameter has been replaced by the particle radius. The theoretical collection efficiency for the cyclone in terms of the impaction parameter can be written:

$$E = \frac{2NRc}{W_i} P_I \qquad (11.19)$$

## 11.5   LIQUID (WET) SCRUBBERS

These control devices are used for the removal of both particulate and gaseous matter. It has been mentioned that centrifugal separators, inertial separators and other devices can also be liquid scrubbers. The common factor is the use of a liquid to contact the gas stream. Liquid scrubbing can be carried out in almost any kind of collector, however, this section is limited to devices whose collection efficiency is primarily dependent on the liquid scrubbing and not the other forces such as inertial, electrostatic and gravitational.

### 11.5.1   SPRAY SYSTEMS

Spray type scrubbers perform like a good rain-storm to remove particulate and absorbable pollutants. These systems consist of such units as spray towers and spray chambers as well as simple waterfalls. The spray systems can use either atomization spray nozzles or centrifugal spray discs as discussed in Chapter IX. Waterfalls provide a sheet of liquid through which the gases move. Additives may be used in the liquid stream to promote chemical reactions, reduce foaming, provide better wetting characteristics or to chemically react with the pollutants. Water alone is a suitable, economical and available liquid for many installations, but other liquids are used to obtain desired absorption and/or chemical reactions.

Operating costs for these devices will depend on the type spray system as well as the scrubbing liquid used. High pressure spray nozzles frequently make it necessary to use pumps to build up the pressure. Centrifugal spray disks may require pumps and meters as well as compressed air or other gas. The liquid in most of these devices falls by gravity and can be either discarded or reused. If reused, it is usually necessary to clean the liquid or at least add some fresh makeup liquid. Cleaning of the scrubbing liquid may require filtering and/or chemical processing to prevent water pollution. Any particulate matter or other chemicals collected should be recovered to provide a dollar return value. Chemicals added to produce precipitates with dissolved pollutants should also be filtered and recovered. It may even be necessary (or economically desirable) to remove absorbed pollutants by extraction processes. Waterfall devices actually consist of simple overflow weirs and have few auxiliary facilities. Although these devices can remove some of the particulates from the effluent gas stream, they are less efficient than inertial separators that operate at a higher pressure drop. No attempt will be made to provide theoretical equations for these devices because of the tremendous variation in design and operating characteristics, but typical spray tower efficiencies are given by Figure 11.16 at the end of this chapter.

## 11.5.2 ABSORPTION (SCRUBBING) TOWERS

The two common types of towers, plate and packed, make extremely effective pollution control devices. They utilize a simple mechanical method of achieving good contacting between the gas and liquid phases to provide favorable overall mass transfer. Towers or columns have been used industrially for many years in chemical operations such as distillation, rectification, absorption and extraction, so they have an advantage of being familiar devices to many engineers. Three general types of towers will be discussed: (1) packed towers, (2) bubble cap towers and (3) perforated plate towers.

### *PACKED TOWERS*

These are cylinders filled with a packing material and supported at the top and the bottom to permit phase separation. The packing in these towers can be nearly anything from broken bottles or pieces of coke to ingeniously prepared shapes such as fitted plates, interlocking saddles, spheres

and cylinders. Packings are made of carbon,
ceramic, glass, plastic, Teflon, stainless steels
and other metals. Some of the commercially available forms of packing are the Beryl saddles,
Intalox saddles, Raschig rings, Pall rings and
Cannon packing.

The requirements for packings are a) provide
good surface contact area, b) give low pressure drop
to gas flow, c) provide even distribution of both
fluid phases (*i.e.*, result in low amounts of
channeling), d) be unreactive with either phase
(unless catalytic influence or chemical reaction is
desired), e) be sturdy enough to support themselves
in the column, f) have abrasion resistance to
prevent being worn away by attrition, g) be economical
for the desired operation (low cost, available and
easily handled).

Figure 11.9 is a schematic of a packer tower
operating with counter-current flow. The liquid
enters at the top of the tower and must be distributed in some manner such as by a spray system,
overflow weirs or a perforated distribution plate.
The liquid should pass uniformly throughout the
packing where it contacts the rising gas. Sometimes
it is necessary to add intermediate packing supports
if the packing is not capable of supporting the
entire column weight of liquid and packing. These
supports can also be used as liquid redistributors
to try to prevent liquid channeling. A packing
support is required at the bottom of the tower.
This screen or plate support grid provides a
receptacle for draining the dirty liquid, and also
distributes the entering gas before the gas starts
upward through the packing. Sometimes it is
necessary to utilize a gas demister at the top of
the tower to prevent entrainment of liquid droplets
in the exit gas stream. This demister can be a
fibrous mesh of metal or other suitable material.
The tower packing must be installed with considerable
care and not simply dumped into the shell. If it
is not added in small quantities and spread evenly
to assure uniform distribution, excessive channeling
will occur during operation. The same tower shown
in Figure 11.9 could be operated co-currently with
little modification by adding the gas at the top
and taking it out at the bottom.

*BUBBLE CAP TOWERS*

A bubble cap tower has no support grids or
packing, but in their place has evenly spaced plates

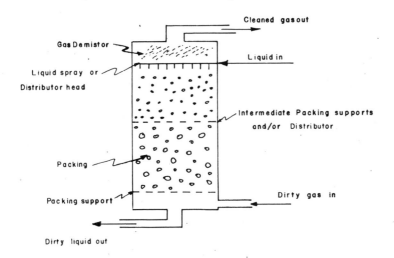

Figure 11.9    Schematic of a Packed Counter-Current
Tower

with specially designed risers and downcomers as
shown in Figure 11.10.  This figure is a sectional
view showing two plates of a typical bubble cap
tower.  The gases enter each plate through the gas
risers which are spaced according to a consistent
geometrical arrangement.  The gas then forces its
way through the liquid and out through slots in or
underneath the bubble caps.  This causes good
turbulence and provides a large surface area for
contact between the gas and liquid phases.
      The liquid on each plate flows to the next lower
plate by means of downcomers which are pipes
containing various levels of overflow weirs.  The
actual height of liquid on each plate depends on
the height and design of the liquid overflows as
well as on the liquid and gas flow rates.  The gas
pressure drop across each plate increases with
liquid height on the plate and gas flow rate.

*PERFORATED (SIEVE) PLATE TOWERS*

      These towers are designed similar to bubble cap
towers except instead of using risers or downcomers,
the plates simply have holes in them.  The number of
holes, their shape and their arrangement on the plates

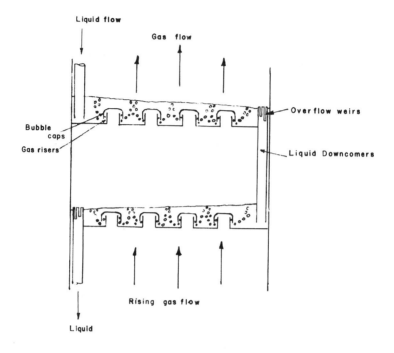

Figure 11.10   Cross Section of Bubble Cap Tower

varies from column to column but are often 1/8"
holes on 3/8" triangular centers in 1/4" thick
plates.   Figure 11.11 shows a section of a sieve
plate tower and the top of a plate.   The rising gas
provides a resistance to liquid flow so that a
liquid head can be maintained on each plate.   The
two phases contact each other as the gas bubbles up
through the liquid as well as when the liquid falls
through the gas.   The liquid level on the plate is
a function of both liquid and gas flow rates.   It
is necessary to have level plates with consistently
sized and arranged holes in order to maintain a
fairly even liquid height on the plate.   Irregularity
would result in channeling or by-passing if all the
gas were to pass through a section having less
liquid resistance to gas flow.   Perforated plate
towers with slotted plates similar to the plate
shown in Figure 11.11 are called Turbogrid towers.

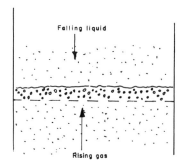

(A)  SIDE  VIEW OF SIEVE PLATE COLUMN(Schematic)

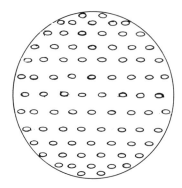

( B) TOP  VIEW OF SIEVE PLATE WITH OVAL SHAPED HOLES

Figure 11.11 Cross Sections of Sieve Plate Tower

11.5.3  ABSORPTION TOWER CAPACITY

It has long been thought that absorber tower capacity was limited by 1) a maximum vapor rate above which liquid would be carried upward by entrainment in the rising vapor concluding effective counter-current separation and 2) a maximum liquid rate above which the column would flood and dump, terminating the operation.  Using these rate limiting restrictions, some tower capacities are given below, but it should be noted that new operational methods discussed in Section 11.5.4 may make it desirable to alter these conditions for counter-current tower operations. The maximum allowable vapor velocity in ft/sec ($v_b$) for bubble cap towers can be estimated (3) using:

$$v_b = K_v \sqrt{\frac{\rho_1 - \rho_g}{\rho_g}} \qquad (11.20)$$

where: $\rho_1$ = liquid density, lb/ft$^3$
$\rho_g$ = gas density, lb/ft$^3$
$K_v^g$ = a constant from Table 11.2

As a general rule, the bubble cap maximum vapor velocity is approximately 3 ft/sec at 1 atm for counter-current operation.

Using the same limiting conditions, packed column flooding velocities occur when the superficial vapor velocity is from 4-6 ft/sec. The superficial velocity is the velocity obtained by dividing volumetric flow rate by the total cross-sectional area of the empty packed column. Flooding has been observed to occur in packed towers when the rising gas velocity is greater than 1.5 times the velocity which results in a pressure drop of 0.5 inches of water per foot of height. These packed column maximum velocities are also valid for counter-current operation.

Tower diameters are readily calculated by:

$$D_T = \sqrt{\frac{4Q}{\pi U_s}} \qquad (11.21)$$

where: $D_T$ = tower diameter, ft
$Q$ = volumetric gas flow rate, ft$^3$/sec
$U_s$ = superficial gas velocity, ft/sec

The height of an absorption tower can be estimated by multiplying the number of overall gas transfer units by the height of a transfer unit as discussed in the next section.

TABLE 11.2   VALUES OF $K_v$ FOR USE IN EQUATION 11.20 (4)

| Plate Spacing, Inches | Liquid Seal Height, Inches | | | |
|---|---|---|---|---|
| | *0.5* | *1.0* | *2.0* | *3.0* |
| 6 | 0.03 | – | – | – |
| 12 | 0.10 | 0.08 | 0.06 | – |
| 24 | 0.18 | 0.17 | 0.16 | 0.15 |
| 36 | 0.20 | 0.19 | 0.19 | 0.18 |

Build-up of deposits and corrosion of the tower can physically change the internal characteristics and, therefore drastically reduce the maximum vapor velocities and required overall height to effect a given separation under specific operating conditions. In counter-current operation, most of the deposits of particulate material occur near the bottom of the tower. Frequently, more scrubbing liquid can be introduced at this section of the tower to reduce potential plugging problems.

## 11.5.4  ABSORPTION TOWER EFFICIENCY

Absorption efficiencies of the packed and plate type towers can be obtained in a similar manner. This can be done by either a mathematical or a graphical procedure, although the graphical procedure is more easily understood and just as accurate. The McCabe-Thiele type graphical procedure consists of stepping off the number of theoretical contact stages or number of theoretical plates (NTP) on a vapor-liquid equilibrium diagram as discussed in Section 10.3.5. The overall absorption efficiency of a plate type tower then becomes:

$$E_o = \frac{NTP}{\text{no. of actual tower plates}} \qquad (11.22)$$

This is the same procedure used in the calculation of distillation efficiency, except in distillation the stillpot counts as one extra actual plate.

There are no plates in a packed tower, so it is necessary to estimate the number of plate equivalents which is the number of transfer units (NTU). This can be done for a packed tower using the Chilton-Coburn Equation:

$$NTU = \int_{y_1}^{y_2} \frac{dy}{y^* - y} \qquad (11.23)$$

where: $y_1$ = vapor composition entering the bottom of the counter-current column, mole fraction

$y_2$ = vapor composition leaving, mole fraction

$y^*$ = equilibrium vapor composition at the point where y is determined, mole fraction

The overall efficiency of the packed tower can then
be found by replacing the number of actual tower
plates in Equation 11.22 with the NTU.
The height equivalent to a theoretical plate
(HETP) or contact stage can be applied to both plate
and packed type towers. It is obtained by dividing
actual tower height by the number of theoretical
stages for a given type separation, tower design
and operation condition. If the HETP value is
available from actual measurements, from the
equipment or packing manufacturer's literature or
pilot plant studies, it can be used to calculate
the approximate height (Z) of a proposed absorber
by:

$$Z = (NTP)(HETP) \qquad (11.24)$$

The value of HETP will vary depending on: type of
plate or packing, column diameter, height of tower
(to a minor extent), mass flow rates and absorbents
and absorbates used.
An alternate procedure for estimating the
height of an absorbing tower is to use the equation:

$$Z = N_G H_G \qquad (11.25)$$

where:  $N_G$ = no. overall gas transfer units

$H_G$ = ht. of a transfer unit

This procedure was mentioned in Section 11.5.3.
Values of $N_G$ can be estimated using the procedure
from Section 10.3.7 and the approximate value of
$H_G$ of 2 feet can be used for some preliminary
estimates. $N_G$ can also be calculated using the
AIChE design manual procedure for bubble cap towers
which gives the equation:

$$N_G =$$

$$\frac{0.776 + 0.116\ h_w - 0.290\ v_g \rho_g^{0.5} + 0.0217\ L/W}{Sc^{0.5}} \qquad (11.26)$$

where:  $h_w$ = overflow weir (liquid downcomer)
height, inches

$v_g$ = gas velocity through risers, ft/sec

L = liquid flow rate, gal/min

W = distance between risers (edge to
edge - See Figure 11.10), ft

$$Sc = \text{Schmidt no. of gas} = \frac{\mu_g}{D'_{AB}\overline{M}} \text{ ,}$$

dimensionless

$\mu_g$ = gas viscosity (usually air = $1.21 \times 10^{-5}$ lb/(sec ft) at SC)

$\rho_g$ = gas density, lb/ft$^3$

$D'_{AB}$ = gas diffusivity (see Equation 10.1), lb moles/(ft hr)

$\overline{M}$ = av. molecular weight of gas phase

The Schmidt number for the gas can be calculated using the diffusivity equation given in Section 10.1. Or as an estimate, the Schmidt number of gases in air at 32°F and 1 atm. ranges from 0.60 for water vapor to 1.284 for sulfur dioxide. Although there are higher Schmidt numbers (for example, n-octane has a value of 2.62) the values for many pollutants are less than 1.28.

Values of $H_G$ can be calculated using the Cornel (5) model which is suitable for Raschig rings and saddle type packings:

$$H_G = \frac{\Psi\, S_c^{0.5}\, D_T^{\,n}}{(L_s f_1 f_2 f_3)^m} \qquad (11.27)$$

where:  $\Psi$ = parameter constant (see Table 11.3)

$m$ = 0.6 for rings

= 0.5 for saddles

$n$ = 1.24 for rings

= 1.11 for saddles

$D_T$ = tower diameter, ft (max. of 2 ft)

$L_s$ = superficial liquid mass velocity, lb/(hr ft$^2$)

$f_1 = \left(\dfrac{\mu_1}{2.42}\right)^{0.16}$ , $\mu_1$ = liquid viscosity, lb/(ft hr)

$f_2 = \left(\dfrac{62.4}{\rho_1}\right)^{1.25}$ , $\rho_1$ = liquid density, lb/ft$^3$

$f_3 = \left(\dfrac{72.8}{T}\right)^{0.8}$ , $T$ = surface tension, dynes/cm

Liquid superficial velocity is calculated using the total cross-section area of the tower, assuming it contains no packing. A maximum tower diameter of

2 feet should be used for $D_T$ in Equation 11.27 even
if the actual tower diameter is greater than 2 feet.

TABLE 11.3   VALUES OF PARAMETER Ψ FOR EQUATION
11.27 (5)

| Gas Phase Ψ for: | Raschig Rings | | Saddles | | |
|---|---|---|---|---|---|
| | 1″ | 2″ | 1″ | 1-1/2″ | 2″ |
| 40% flood | 110 | 210 | 60 | 80 | 95 |
| 60% flood | 105 | 210 | 60 | 80 | 95 |
| 80% flood | 80 | 210 | 60 | 80 | 95 |

Typical gas absorption efficiencies and
operating characteristics of the common types of
absorption towers are listed for comparison in
Table 11.4. All but the cycled tower are operating
below the standard maximum vapor velocity and
maximum liquid thruput limitations.

Contact efficiencies for absorption and other
mass transfer operations can be vastly improved by
a practice called controlled cycling. Controlled
cycling, which is not to be confused with pulsation,
can be used on counter-current operations that have
stepwise type separation as, for example, in the
plate type towers. The operating cycle consists of
two parts: 1) a vapor flow period when the vapor
flows upward through the column while the liquid
remains stationary on each plate, 2) a liquid flow
period when no vapor flows and liquid drains to the
next lower plate. An absorber that is cycled at a
rate so as to permit all the liquid on a tower plate
to be transferred before restarting the cycle gives
the highest absorber efficiency for a given set of
conditions (6). Under these conditions, the cycle
times vary from 20 seconds to 2 minutes and vapor
flows about 80% of the time. Mathematical studies
show that efficiencies of 200% and even higher are
theoretically possible (6,7).

Thus far we have considered collection efficiency
of absorbers in relation to the removal of gaseous
pollutants. Overall efficiency can also be
calculated for removal of particulate type pollutants
by measurements made on the inlet and outlet gas
streams. Alternate methods for obtaining overall
particulate collection efficiency in absorbers by
relating theoretical collection and capture
equations have been made. See the presentation by
Calvert (2).

TABLE 11.4  WET SCRUBBER TOWERS (COLUMNS)

| Type | Plate Liquid Thruput, Relative | | Efficiency, % | Overall Pressure Drop, inches H2O* | Plate Spacing, inches | Approximate Relative Initial Cost** |
|------|------|------|------|------|------|------|
| | Low | Medium | | | | |
| Bubble cap | 1 | 1 | 60-80 | 16.0 | 16-32 | 1 |
| Sieve plate | 1.2 | 2.0 | 60-80 | 8.0 | 8-32 | 2/3 |
| Packed | 0.7 | 1.3 | --- | --- | --- | 1-1/2 (with ceramic rings) |
| Cycled tower, sieve | 1.2 | 1.4 | 80-96 | (variable) | (variable) | 2/3 |

\* For 5 plate or 20 ft packed column

\*\* Same materials of construction

11.5.5   SORPTION

The process called sorption, which originates when the processes of absorption and adsorption are combined, can be used effectively in air pollution control. Several studies have been made at the Center for Air Environment Studies at the Pennsylvania State University [the last study was reported by Morgan (6)] showing that systems such as carbon in water improve the removal of $SO_2$ from air. The sorption process consists of 1) absorption-- where the gas is transported to the gas film, the molecules diffuse through the gas film then through the liquid film, and finally bulk transport moves the absorbed material to the solid surface and 2) adsorption--where the absorbed gas is removed from the liquid when the gas passes into the pores of the solid. By this removal of absorbed gas, sorption increases the concentration gradient driving force. Even greater efficiencies are possible by cycling the sorption process.

11.5.6   ADSORBERS

Commercially available adsorbers composed of thin beds of adsorbent on a support material are able to handle up to 2000 $ft^3$/min of gas. These adsorbers may be either stationary or fluidized beds and are either regenerative or non-regenerative systems. Non-regenerative systems find use in applications where concentrations are extremely low (for example, odors at less than 2 ppm) or for small laboratory type installations. Adsorbers must be designed to meet the following requirements: 1) retention of the gases to permit sufficient contact time between the gaseous and solid phases is required, 2) the system capacity must exceed the breakthrough point which occurs when excessive pollution leaves with the exit gas (this occurs somewhere before the adsorbent is saturated, frequently at 75 to 80% saturation, but sometimes as low as 15% of saturation, depending on materials and operating conditions), 3) the system should give low gas flow resistance, 4) uniform distri- bution and packing is necessary to prevent gas channeling, 5) precleaners are needed to remove particulate matter, and 6) a spare adsorber is required in parallel for continuous systems to permit regeneration and/or replacing of the spent system.

The cycle time in hours for two different types of commercial adsorbers shows kinds of operating extremes that can be encountered. In this example, each of these units is offstream as long as it is onstream. The pressure swing adsorber has a total cycle time of 48 hours and the temperature swing solvent recovery adsorber has a cycle time of 4 hours.

|  | *High Pressure Gas Dryer* | *Organic Solvent Recovery Unit* |
|---|---|---|
| Hours onstream | 24 | 2 |
| Hours offstream: |  |  |
| Purge at 1 atm | 2 | - |
| Hot gas drying | 10 | - |
| Steam stripping | - | 0.75 |
| Hot gas drying | - | 0.33 |
| Cold gas cooling | 5 | 0.42 |
| Standby | 7 | 0.50 |
| Offstream total | 24 | 2.00 |

Turk (8) presents an equation for the typical average retention time required to completely adsorb an organic vapor from air:

$$t = \frac{1.29 \times 10^6 \ W}{Q \ y_i \ \overline{M}} \qquad (11.28)$$

where: $t$ = duration of adsorbent service before saturation breakthrough, hr
$W$ = wt. of adsorbent, lb
$Q$ = volumetric air flow rate, ft³/min
$y_i$ = vapor inlet conc. in the air, ppm
$\overline{M}$ = av. molecular wt. of adsorbed vapors

When there is a mixture of several organic vapors, the value for M can be obtained using the chain rule:

$$\overline{M} = n_A M_A + n_B M_B + \dots n_x M_x \qquad (11.29)$$

where: $n_x$ = no. moles of component x
$M_x$ = molecular weight of component x

## 11.5.7 VENTURI SCRUBBERS

The Venturi scrubber is a high energy inertial impaction atomizing scrubber used for both gas and

particulate control.  This high efficiency device
is about 9916% efficient for the removal of 5
micron diameter particles and is best suited for
applications where particles are small (0.5-5
microns).  The scrubber construction is similar to
a Venturi flow meter with a 25° converging and a
7° diverging section as shown in Figure 11.12.  The
Venturi scrubber has a 4:1 area reduction between
the inlet and the throat areas (the throat is the
narrowest section of the device--in a Venturi meter
the throat area may be 1/16 the inlet area).  The
Venturi can be either circular or rectangular.
The low angle of divergence enables a high pressure
recovery--up to 90% ΔP recovered in a dry system
and 25-70% in a wet scrubber.  The scrubber throat
length should be from 1/4 to 1/2 the upstream
diameter.

Gas entering the Venturi is smoothly accelerated
in the converger until it reaches a maximum velocity
of from 150 to 600 ft/sec in the throat.  This converts
the static pressure head to a kinetic energy head
and requires from 20 to 35 inches of water pressure
drop.  Scrubbing liquid is atomized by the high
velocity gas stream to produce droplet particles which
act as targets for impaction type collection.  Either
drop-type or cloud-type atomization can be made to
occur as discussed in Section 9.7.  When water is
used as the scrubbing liquid, the rate may range from
5 to 40 gallons of water per thousand ft$^3$ of gas.
The scrubbing liquid can be introduced by various
methods and locations.  The two most common injection
locations are at the throat (the recommended location)
and the entrance of the converger.  Liquid introduced

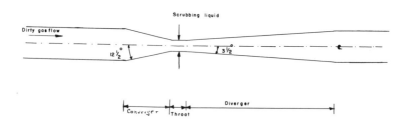

Figure 11.12   Venturi Scrubber

into the throat can be directed perpendicularly
into the moving gas stream or it can be directed
into or away from the moving gas stream. Perpendicular injection of liquid at the throat into the
gas stream appears to be both most economical and
most efficient. In all other procedures, it is not
possible to achieve as uniform a liquid distribution
across the throat.

Energy from the gas stream is used to atomize
the liquid. For cloud-type atomization, this
accounts for only about 6% of the available gas
stream energy, and for drop-type atomization this
amounts to only about 0.5% (based on the work
required to form the increased surface area). Good
atomization is desirable to produce sufficient
inertial impaction targets for particulate matter
collection and to produce a high surface area for
absorption. It is also desirable to keep the
droplet acceleration rate as low as possible to
provide the greatest velocity difference between
the particle being collected and the droplet target.

Nozzles greater than 1 mm inside diameter
permit cloud-type atomization to take place. This
produces extremely fine particles which appear to
have a lower than normal acceleration rate. Tests
completed at this time show no advantage in using
cloud-type atomization. However, it seems logical
that gas absorption should be greatly increased,
and if the particulate matter is sufficiently
small, greater particulate collection efficiency
should also be possible using cloud-type
atomization.

Ingebo (9) has shown that acceleration of
droplets is a function of droplet diameter. In a
constant velocity gas stream, the distance a
droplet travels until it reaches 90% of the gas
velocity is proportional to the particle diameter:

$$\frac{S_1}{S_2} = \frac{2d_1}{d_2} \qquad (11.30)$$

where:  $S$ = distance to accelerate a droplet to
90% of the gas velocity
$d$ = droplet diameter (use $d_1 > d_2$)

A one hundred micron droplet, such as would be
produced from a spray nozzle pressurized at 100 psi,
would reach 90% of the gas velocity when accelerated
by a 100 ft/sec gas stream in 16 inches if it had
no initial forward velocity. It could then be
expected that a 30 micron droplet, which could be

produced by a spray system utilizing a gas stream
with a liquid-to-gas spray volume ratio of 0.001,
would be accelerated to 90 ft/sec in only 3 inches
if it also had no initial forward velocity. This
indicates that most of the gas cleaning in a
Venturi scrubber must be done within the first
several inches of the diverger.

A high gas stream velocity is required in order
to atomize the injected liquid by pneumatic two-
fluid atomization. Figure 9.14 shows minimum
critical velocities required for atomization of
water by an air stream for various size inlet
nozzles to produce cloud-type atomization. It is
recommended, though, that no throat velocity of
less than 150 ft/sec be used. It is important that
the liquid be injected uniformly across the throat
of the scrubber. Use of too large a liquid inlet
nozzle is to be guarded against as it would not
permit the water to be projected into the moving
gas stream, where it must be in order to be
atomized. The scrubber should be designed for the
ratio of scrubber throat diameter (or height) to
water nozzle diameter to be approximately 9:1 (10).

Particulate matter in the gas travels at
essentially the same speed as the gas in the
Venturi scrubber and is captured on the slow
moving freshly atomized liquid droplets by inertial
impaction. The exact mechanism for particulate
collection is unknown so theoretical collection
equations cannot be given  but the fact that a
Venturi scrubber approaches 100% efficiency for
particles greater than 1.5 microns in diameter
makes it valuable for air pollution control. Gas
absorption efficiency can be estimated using the
rate of gas absorption equation:

$$r_a = (C-C^*)\left(\frac{D'_{AB}}{d'}\right)(Sh) \qquad (11.31)$$

where:  $C$ = concentration in gas, mole fraction
$C^*$ = concentration if gas were in
equilibrium with liquid at location
where C is taken, mole fraction
$D'_{AB}$ = gas diffusivity (see Equation 10.1)
lb moles/(ft hr)
$r_a$ = rate of gas absorption, lb moles/
(ft$^2$ hr)
$d'$ = scrubbing droplet diameter, ft
$Sh$ = drop Sherwood No., dimensionless
= $2 + 0.6 (Sc)^{1.3} (Re)^{\frac{1}{2}}$
= 2 to 10
$Sc$ = Schmidt No. (see Equation 11.26),
dimensionless

Re = drop Reynolds no., dimensionless

It is difficult to use equations of this type
because droplet diameters vary so widely and
concentrations change so rapidly in a very short
distance. Venturi absorption is equivalent to 1 to
1-1/2 theoretical plates.
Venturi conditions can be summarized as follows:

Gas flow:  200 to >145,000 cfm
Gas velocity in throat:  150 to 160 ft/sec
Liquid to gas ratio:  5 to 40 gallons of water/
                      1000 ft gas
Pressure drop:  20 to 32 inches of water
Temperature of inlet gas:  up to 500+°F

A typical particulate collection efficiency curve
for a standard Venturi scrubber using drop-type
atomization is given in Figure 11.17.

## 11.6 FILTERS

Fibers of filter media act as inertial impaction
targets to collect particulate matter. Fabrics have
the additional advantage of containing hair-like
appendages (nap) on the fibers which act to improve
the collection efficiency of the filter. Fiber
filters usually do not have sufficient strength to
support themselves in a rigid shape, so they are
either supported on a backing or made into shapes
such as long cylindrical bags and suspended from
hangers.
Filter bags are usually made at least 4 inches
or larger in diameter and generally up to 40 feet
in length. They can be hung vertically in banks
of several dozen bags in one of the  several
standard designs of bag houses. In one type bag
house, dirty gas is sucked into the bag through an
open bottom end, then passed out through the sides
of the bag leaving the dirt inside. In another
system, the dirty gas is sucked through the bags
leaving the dirt on the outside. The cleaned gases
pass upward through the bags and exit through a
hole in the top.
Cleaning of the bags is an important art, with
at least five systems commercially available. The
first system necessitates stopping gas flow to the
filter in contrast to the usual system in which the
gases are diverted to an alternate bag or bags.
When the bag house has been isolated, the bags are
literally shaken by a mechanical or manual device.

In a second system, the gases continue to flow, but they are directed to various sections of the bag house.  When certain bags are not receiving dirty gas, they are collapsed by a reverse flow of clean air which pulses into the bag to knock the dirt off and carry it away.

A third system utilizes a ring-type sparger which passes from top to bottom of the bag while the bag is still in operation.  The high velocity air jet moves against the bag from the ring and loosens the dirt so that it falls downward by gravity.  If the loosened dirt is forced back against a lower section of the bag by the flow of dirty gas, the jet will free it again the next time it moves on its downward path.

Another system can be cleaned during operation by impressing a sound energy wave into the system to vibrate the bags.  Sonic energy impulse couplers can be used to shake all the bags simultaneously during operation.

The fifth system uses combinations of several of the above systems.  In all cases, the dirt collected from the bags falls to the bottom of the bag house after which it can be removed by gravity through a blast gate or rotary air lock valve.

## 11.6.1  FILTER FABRICS

Fibers are made into cloth by either weaving yarns of the fibers or by pressing the fibers directly into felts.  There are three basic methods for weaving filter fibers.  They are called plain, satin and twill, with satin and twill being more useful for gas filtration purposes.  The plain weave is a simple "one up" and "one down" type construction.  It has the characteristic of being a tight, relatively impermeable weave which causes high resistance to gas flow, although it can be useful for ultra-fine filtering.  This weave is shown in Figure 11.13a.

The satin weave, as shown in Figure 11.13b, has fewer interlacings with more widely spaced fibers and an irregular appearing surface.  The increased porosity gives less gas flow resistance.  In cotton, this type of weave is commonly known as "sateen".  The third weave is the twill, recognizable by a sharp diagonal line which is the result of fewer interlacings.  The twill is more porous than the plain weave though not as porous as the satine.  Figure 11.13c shows three of the many twill structures (the ones shown are: 2 over, 1 under;

2 over, 2 under; and 3 over, 1 under). Fabrics can
also be layered if high particulate collection
efficiency is desired. This obviously results in
increased pressure drop and can also cause cleaning
problems.

Felts, made by pressing the fibers to lock them
together mechanically, can be made of natural or
synthetic fibers, or of the two combined. This
type fabric offers greater resistance to air flow
because of the closely packed material, but for the
same reason makes a very efficient filter cloth.
A disadvantage of felt filter material is that it
is difficult to clean.

Numerous materials of construction are available
for custom-making filter fabrics for specific
applications as indicated in Table 11.5. When
choosing material, it is important to consider the

(A)   PLAIN WEAVE (Enlarged cross section)

(B) SATIN WEAVE (Enlarged cross section)

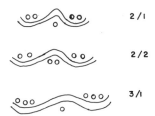

(C)   TWILL   WEAVE (Enlarged cross section)

Figure 11.13   Three Types of Filter Fabrics

safety factor. Flammable or explosive dusts could
be ignited by a static spark discharge, for example.
Some fabrics have a higher tendency to build up
static charges than others and are so noted in
Table 11.5 by the magnitude of the positive or
negative value listed. The combination of static
charges on the dust and/or bags can also create
operating conditions that make it impracticable to
control some dusts by filtration. The data for the
various properties shown in Table 11.5 are obtained
from manufacturers' literature with the comparisons
supplied by this author except as noted.

Special fibers have been developed to adsorb
gases as well as collect particulates. An example
is dimethyl aminopropyl maleimide copolymer which,
when co-spun with polypropylene, can be used to
adsorb sulfur dioxide at 55°C. This material can
even be reactivated by treating with alkaline
solutions. It is expected that more fibers of this
type will be developed and used.

Glass and steel have the highest usable
temperatures but are subject to other weaknesses.
Glass will not stand much flexing and can wear out
quickly, break and require early replacement if not
used carefully; the use of the glass bags does not
necessarily result in abnormally high overall annual
costs because the life of these bags in continuous
service can be more than 5 years. Steel is subject
to oxidation and possible chemical attack. Graphite
fibers have specific application but are extremely
expensive, currently selling for $350-$500/lb. It
is estimated that by 1975, the price may still be
about $100/lb, despite increased production.

## 11.6.2 FILTER EFFICIENCY AND CAPACITY

The efficiency of a filter bed increases with
the pressure drop (as long as adequate air velocities
are maintained for inertial impaction of particles)
but capacity decreases with increased pressure drop.
A filter is not very efficient until it has been
"precoated" by allowing an initial layer of dirt to
accumulate or unless a layer of special material
has been deposited on the upstream side of the
filter. Overall pressure drop is: the sum of the
pressure drop due to the filter itself, plus the
pressure drop due to the particles accumulating on
the filter media.

$$\Delta P_o = \Delta P_F + \Delta P_P \qquad (11.32)$$

TABLE 11.5  COMPARATIVE TABLE OF FILTER FABRIC PROPERTIES

| Material | Temp. Limit | Strength | Resistance to Acid | Alkali | Organic Solvents | Flex & Abrasion | Dust Release | Flame Resistance | Relative Static Generation (11) | Cost Ratio to Cotton* |
|---|---|---|---|---|---|---|---|---|---|---|
| Asbestos | 500°F | P | G | G | G | P | F | E | / | 5.0 |
| Carbon (Graphitized) | 500°F | P | F | G | G | P | G | P | / | 30.0 |
| Cotton | 180°F | G | P | F | G | G | F | P | +6 | 1.0 |
| Olefin Treated Cotton | 250°F | E | G | G | G | G | G | G | / | 2.7 |
| Dacron Polyester | 280°F | E | G | G | E | E | E | E | 0 | 2.7 |
| Dynel | 210°F | G | E | E | F | G | G | E | -4 | 3.2 |
| Glass (spun yarn type) | 750°F | F | E | P | E | P | F | E | / | 6.0 |
| Glass (cont. filament silicone treated) | 500°F | F | F | F | E | F | E | E | +15 | 2.2 |
| Nylon ("Nomex") | 425°F | E | F | G | E | E | E | E | +10 | 8.5 |
| Nylon (6,6) | 200°F | E | F | G | E | E | E | E | +10 | 2.5 |
| Orlon Acrylic | 275°F | G | G | F | E | G | G | P | +4 | 2.7 |
| Paper | 180°F | P | P | P | G | P | F | P | / | 0.5 |
| Polyethylene | 250°F | G | G | G | E | G | G | P | -20 | / |
| Polypropylene | 225°F | E | E | E | G | E | G | P | -13 | 1.5 |
| Polyvinyl acetate | 250°F | G | G | G | P | G | E | P | / | / |
| Steel | 800°F | G | E | F | F | G | G | E | -10 | / |
| Teflon | 500°F | G | E | E | E | P | G | E | / | 25.0 |
| Wool | 210°F | G | G | F | G | G | F | P | +11 to +20 | 3.7 |

* Varies with type of weave and weight of fiber.  E = Excellent, G = Good, F = Fair, and P = Poor

where:   $\Delta P_o$ = overall pressure drop
$\Delta P_F$ = filter fabric pressure drop
$\Delta P_p$ = particle pressure drop

Values for the filter pressure drop under various flow conditions can be obtained from manufacturers' literature or from experimental data. The resistance due to the precoat which is obtained after the dust has been removed several times becomes a constant that should be included in the fabric pressure drop. Pressure drop caused by the collected particles increases with dust build-up. This can be estimated using the Kozeny Carmin differential equation:

$$\frac{d(\Delta P_p)}{d\ m} = \frac{180\ \mu\ v}{g\ \rho_p(d')^2} \left(\frac{1-\Theta}{\Theta^3}\right) \qquad (11.33)$$

where:   $\mu$ = gas viscosity, lb/(ft sec)
$v$ = superficial gas velocity, ft/sec
$\rho_p$ = particle density, lb/ft$^3$
$d'$ = particle projected mean diameter, ft
$m$ = dust loading on filter, lb/in$^2$
$\Theta$ = porosity, pore area/total area
$\Delta P_p$ = lb/in$^2$

The numerical value of the porosity is always less than 1 and can be determined by experiment for the particular fiber, fiber construction and type of dust.

Filtration efficiency for a given particle passing through a bed of cylindrical fibers is:

$$E = 1 - \exp\ [-\eta\ S] \qquad (11.34)$$

where:   $\eta$ = impaction target efficiency (see Figure 11.14)
$S$ = total projected area of all obstacles in the bed divided by the cross-section of filter normal to gas flow

The significance of S is that it is the number of times a particle has a chance to be removed from the gas stream. Another way of determining the value of S is to divide the net projected area of all layers by the cross section area of one layer of filter. Values for the fractional impaction efficiency can be obtained from Figure 11.14 which

is a plot of efficiency versus the dimensionless
parameter:

$$\frac{D_c \, g}{v_p \, v_s}$$

where:  $D_c$ = diameter of fiber
$v_p$ = particle velocity
$v_s$ = Stokes' settling velocity

As the particle size decreases, the impaction
efficiency can decrease as indicated by Figure
11.18, a particulate collection efficiency curve
for a typical filter fabric. As particle diameter
continues to decrease, the fabric filtration
efficiency can increase as shown in Figure 11.18,
but for these extremely small particles the
collection is a function of diffusion rather than
impaction.

Optimum filter operation appears to be obtained
when the *superficial velocity* of the gas passing
through the cloth is *1-10 ft/min* (12) with ranges
of 3-6 ft/min common for fine dusts and 6-10 ft/min
for coarse dusts. Large, easily filtered dusts can
be filtered at velocities up to 25 ft/min (13).
Frequently, pre-cooling is utilized to lower the
gas temperature so that a more economical choice of
cloth can be utilized. Care must be taken when
pre-cooling gases to be sure that the dew point
will not be reached in the filter. If this happens,
vapors will condense and clog the pores of the
cloth. Heat exchangers utilized to pre-cool the gases
must be frequently inspected to insure that thermal
precipitation has not resulted in blockage due to
particle deposition. An alternate method of reducing
gas temperature is to dilute the hot gas with ambient
air. Dilution can also improve filtration when the
gas to be cleaned has an extremely high dust
loading. The additional air permits a higher
filtration velocity (up to 10 ft/min) which aids in
inertial impaction collection. Gas absorption and
adsorption using filters impregnated with chemicals
is being attempted. This could prove to be an
effective means for removal of pollutants such as
odors which are present in very low concentrations.

## 11.7  ELECTROSTATIC PRECIPITATORS

Electrostatic precipitators are used for
particulate pollution control. They may be used

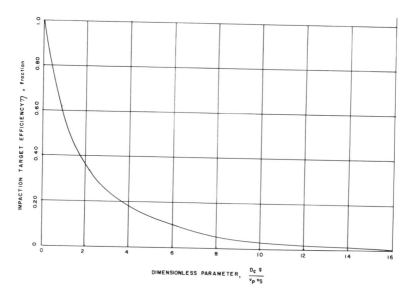

Figure 11.14 Filtration Target Efficiency
(cylindrical Fibers)

either dry or wet (irrigated). The wet type uses
a liquid film (often clean, demineralized water) to
remove the solids from the collector. The dry
collectors must be cleaned by rapping, scraping,
blowing out, or by inducing sonic energy vibrations.
Care is required to prevent re-entrainment of the
dust being removed in the dry system while the
disadvantage in the wet system is that there is a
greater pressure drop to gas flow which increases
operating costs. Electrostatic precipitators have
the advantage of operating with very low pressure
drops across the system, while their main disadvantage
is that continuous cleaning is necessary. It is
also necessary to assure that the dust in the inlet
gas stream is evenly distributed. This is achieved
by the use of vanes or by using a perforated plate
at the inlet of the precipitator.

Precipitators may be arranged in parallel for
greater thruput or in series forgreater efficiency.
The problem with parallel systems is that it is
difficult to attain equal dust distribution for all
systems. Even though electrostatic precipitators
are highly efficient, installations consisting of
up to 10 sections in series are sometimes required
to obtain the desired overall collection efficiency.

Using generalized efficiency curves for particulate collection in irrigated and dry electrostatic precipitators as given in Figure 11.16, typical overall efficiencies can be estimated by the procedure of Section 11.1 (see also Section 9.5.4).

Ramsdell (14) presents a method to determine the required electrostatic precipitator size based on data obtained from coal fired boilers. The formula for overall collection efficiency of an electrostatic precipitator is expressed as:

$$E_O = 1 - \exp\left[-\frac{A_s}{Q} W\right] \qquad (11.35)$$

where:  $E_O$ = overall collection efficiency, percent
$A_s$ = collector surface
$Q$ = volumetric flow rate
$W$ = particle velocity constant related to useful electrical power and coal sulfur content

The Ramsdell Equation for overall efficiency is:

$$E_O = 1 - \exp\left[-RB\right] \qquad (11.36)$$

where:  $R$ = a constant dependent upon coal sulfur content
$B$ = number of active bus sections per 100,000 cfm gas

An active bus section is intended to mean a seperately energized precipitator section. For example, one bus section could be either one section powered by a single rectifier or two sections powered by a double half wave rectifier. An electrical disturbance in either of these would not affect any other sections.

The values of the constants reported by Ramsdell for precipitators operating at 300°F and using medium and low sulfur coal are:

| | *Coal Sulfur Content* | | |
| *Constant* | *0.5%* | *1.0%* | *2.75%* |
|---|---|---|---|
| R | 0.9 | 1.0 | 2.1 |
| W | 1.4 | 1.6 | 3.3 |

Precipitators operating "hot" (about 700°F) on low sulfur coal boilers may have the same fly ash resistivity and therefore the same values for these constants, but this should be checked. Decreased sulfur in coal results in decreased resistivity (see Figure 9.10) and therefore decreased precipitator efficiency at all temperatures.

Knowing the desired overall collection efficiency, coal sulfur content, and gas flow rate, the electrostatic precipitator can be sized for collector area and bus sections using Equations 11.35 and 11.36, remembering that low velocities are still required to permit collection and to prevent re-entrainment.

Operating conditions for electrostatic precipitators are:

Gas flow--1 cfm to $2 \times 10^6$ cfm
Gas temperature--up to 1200°F
Gas pressure--up to 150 psi
Gas velocity--3 to 15 ft/sec (up to 50 ft/sec in special units)
Pressure drop--0.1 to 0.5 inches of water per section
Particles removed--0.1 to 200 microns
Particle inlet concentration--$0.15 \times 10^{-3}$ to 15 lb/1,000 ft$^3$
Treatment sequence--1 to 10 sections in series
Power supply--50,000 to 70,000 volts dc (up to 100,000 volts in some units)
Discharge electrodes--up to 0.109 inch diameter coppered steel wires

## 11.8  OTHER CONSIDERATIONS

There are many other air pollution control devices not mentioned specifically in this chapter. These are the more specialized devices which have less generalized applicability. For further information on these special devices, the reader is referred to other literature. For example, incinerators, which have extensive use, are described in detail by ASME, ASTM, state and local (metropolitan) publications. Sonic precipitators have been used infrequently so far but are discussed by Strauss (13) and Mednikov (15).

Emissions from collectors can still contain air pollutants that have not been recovered. This material can be released into the atmosphere if it does not exceed the emission factors discussed in Chapter II. The total combined emissions in a given location create the ambient air quality. Specification of these emission rate limits for a given geographical area should be a function of six factors:  1) type of operation (for example, whether it is a steel mill or chemical plant with higher emissions per square foot than would be

typical from auto assembly line operations); 2) type
of community (residential, commercial or industrial
metropolitan or rural); 3) size of operation (this
can be related to capital investment); 4) efficiency
of collector (varies with types of pollutants);
5) ambient air quality standards (how clean the
air should be) and 6) meteorological and geographical considerations (history of inversions, winds
and other factors plus physical location which
could create added stresses, such as valleys, hills,
lakes, and roughness of terrain).

Various emission formulas are in use. For
example, in Illinois the amount of allowable particula
particulate emissions (E) in pounds per hour, from
new process emission sources where the input raw
material process weights range up to 450 tons per
hour, are determined by:

$$E = 25.4 \ P^{0.534} \qquad (11.37)$$

where:  P = process weight rate in tons/hour

For operations which use 450 tons per hour or more
input process material, the particulate emissions
are limited to:

$$E = 2.48P^{0.16} \qquad (11.38)$$

In control, design, and emission considerations,
it should be remembered that the stack has an
important function.  Some relevant factors to
consider are:  1) the height of a stack should be
at least 2 to 2.5 times higher than surrounding
structures or terrain; 2) gas exit velocity should
be greater than 60 ft/sec to avoid creating downwash
of the gases due to the turbulent wake of the stack;
3) stacks of diameter less than five feet and
height less than200 feet will permit gases to
strike the gound before appreciable diffusion
occurs--stacks higher than this will more closely
follow the diffusion equations presented in
Chapter III; and 4) ground concentrations of the
pollutants released from a stack decrease in
relation to the square of the stack height (see
Chapter VIII).

## 11.9 GENERALIZED PARTICULATE COLLECTION EFFICIENCY CURVES

Here are particulate collection efficiency curves
for several relatively standard control devices,

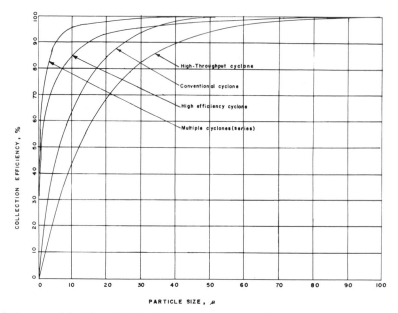

PARTICLE SIZE, μ

Figure 11.15   Efficiency Curves of Cyclone Dust
Collectors (Efficiency versus Particle
Diameter)

presented together for convenience and comparison.
Figure 11.15 shows the efficiency curves for
several cyclone types discussed in Section 11.3.
The spray tower and orifice scrubber of Figure 11.16
are both spray systems as discussed in Section 11.5.1.
The electrostatic precipitators in Figure 11.16 were
discussed in Section 11.7.  Figure 11.17 includes
the Venturi scrubber which was discussed in Section
Section 11.5.7 as well as the impingement scrubber
wich is a form of spray system discussed in Section
11.5.1; and the wet dynamic scrubber is a centrifugal
device which could be classified in both Sections
11.3 and 11.5.  The filter efficiency curve (Figure
11.18) applies to Section 11.6.

QUESTIONS FOR DISCUSSION

<u>1.</u>  Explain what the term (1-E) is related to.
<u>2.</u>  What general comments can be made relating to
particulate and gaseous collection devices?
<u>3.</u>  Can you see how the inertial impaction parameter
is of any value?
<u>4.</u>  How is it possible to tell when an absorber is
flooded?

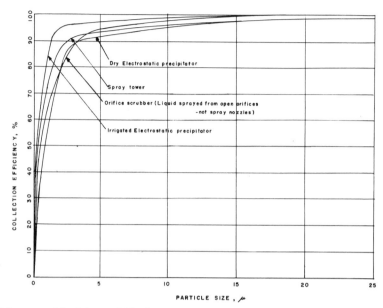

Figure 11.16    Efficiency Curves of Scrubber and
                Precipitator Type of Dust Collectors
                (Efficiency versus Particle Diameter)

5. Can you tell when an absorber is dumping?
6. What is the relation between theoretical contact
stages and actual contact stages in a bubble cap,
sieve plate and packed absorber?
7. Define absorption, adsorption and sorption.
8. What is a McCabe-Thiel diagram?
9. What is the significance of terms such as "$N_G$"
and "$H_G$"?
10. Determine how absorption efficiency can be
theoretically established in a Venturi scrubber.
11. Sketch the flow diagram for a bag house
including recovery of the collected dust.
12. Describe what the top view of the three
different types of filter weaves discussed would
look like.

PROBLEMS

11.1 Estimate the overall efficiency for the
collection of Mgo $d_{50}$ = 6µ by wt. and σ = 2.0  by:
a) a single conventional cyclone separator, b) a
Venturi scrubber, c) a fiber filter, and d) a dry
electrostatic precipitator.  List the devices in the
order of their efficiency.  Use Figures 11.15-11.18.

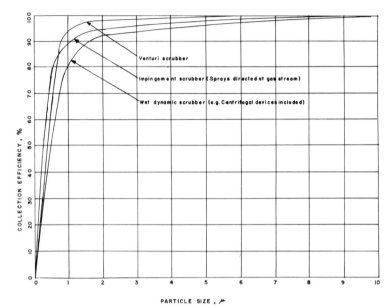

Figure 11.17   Efficiency Curves of Some Wet Scrubber
Type of Dust Collectors (Efficiency
versus Particle Diameter)

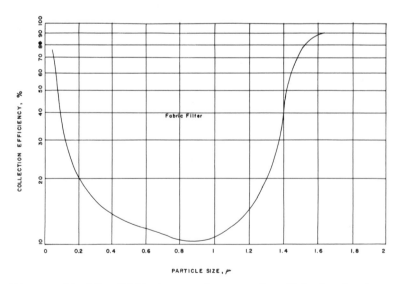

Figure 11.18   Efficiency Curve of a Fabric Dust
Collector (Efficiency versus Particle
Diameter)

11.2 Calculate the diameter of the smallest particle
that can be captured with 100% efficiency in a
simple chamber 20 ft long, 6 ft wide and with an
outlet duct height of 2 ft. The particle density
is 1.8 $g/cm^3$ and the gas is moving at a velocity
of 8 ft/sec in the inlet duct.
11.3 Calculate the overall efficiency of the settling
chamber of Problem 11.2 on fly ash produced by
burning polverized coal.
11.4 Design a high thruput cyclone to handle 6,000
cfm of inlet gas.
11.5 What is the minimum sized particle that can
be collected with 100% efficiency from the cyclone
designed for Problem 11.4 if the inlet gas stream
contains 1.0 grain/$ft^3$ of pulverized coal fly ash
at 650°F?
11.6 Estimate the pressure drop that would be
experienced in the cyclone for conditions of
Problem 11.5.
11.7 Estimate the size of a dry packed bed to
handle 4,500 cfm of BOF gas containing 0.5 grains/
$ft^3$ of iron oxide fumes at 1,000°F if the superficial
gas velocity is 15 ft/sec maximum. If the bed is
20 ft high, what is the overall efficiency if it
is packed with one inch Raschig rings? Assume the
iron oxide fume has by wt. a $d_{50}$ of 1.5μ and a σ
of 1.8. Comment on the use of this column for this
purpose. Should ceramic or plastic packing be used?
11.8 The gases from a coal burning power plant are
to be passed through a counter-current absorber for
removal of both particulates and $SO_2$. Coal at a
rate of 50 ton/day containing 3-1/2% sulfur is
burned in the normal manner. The concentration of
$SO_2$ leaving the absorber is to be a maximum of 50
ppm. a) If the tower is a bubble cap with a plate
spacing of 12 inches, what is the maximum vapor
velocity? b) What is the diameter of the tower in
feet? c) If the slope of the operating line is
0.9 and a $Na_2SO_3$ solution is the absorbent, how
many moles/minute of solution are required? d)
How many theoretical plates are required for this
separation if a pseudo equilibrium curve for these
conditions is y = 0.8x? e) If this is a gas phase
controlling absorption, what is the estimated
overall efficiency? (Neglect the fact that ammonia
absorption by water is liquid phase controlling.)
f) How many actual plates are required for this
tower? g) Calculate an estimated height for this
tower using two different methods.

11.9   Repeat Problem 11.8 this time using 4,000 cfm
of inlet gas containing 0.05 moles of ammonia per
mole of inlet gas entering the absorber.  Concentration of ammonia leaving the absorber should be no
more than 100 ppm.  The operating line for this
absorption has a slope of 1.0 and water is the
absorbent.

11.10  Hydrogen sulfide is to be adsorbed from a
waste gas stream.  The initial concentration is
200 ppm and it is desired to decrease the
concentration of hydrogen sulfide to the limit of
odor detection (0.47ppb).  a) What types of
adsorbents could be used?  b) Assume the adsorbent
is activated alumina beads.  How many pounds of this
adsorbent would be required to clean up a gas stream
with a volumetric flow rate of 800 ft$^3$/min?  (Assume
the adsorber will be onstream for a maximum of two
hours and that breakthrough occurs at 40% of saturation.)  c) What is the bed diameter and height if
the maximum superficial gas velocity is 1.5 ft/sec?

11.11  Estimate the absorption efficiency of a
Venturi scrubber on sulfur dioxide absorption.
Assume that drop-type atomization is used and that
the throat velocity is 200 ft/sec with a liquid to
gas ratio of 5 gallons of water/1000 ft$^3$ gas.  The
inlet concentration of $SO_2$ is 375 ppm.  Size the
unit and determine what exit $SO_2$ concentration
would exist if the gas flow rate is 2,000 cfm.

11.12  The metal fumes from an open hearth furnace
are to be removed by a glass fiber dust collector.
The gas contains 1 grain/ft$^3$ of dust and has a flow
rate of 1000 cfm at 450°F.  The porosity of the
dust is 0.68.  Assume the fumes have a $d_{50}$ of 1.5μ
and a σ of 1.6 by weight and their density is
2.5 g/cc.  The initial pressure drop of the filter
system when first put into service is 10 inches of
water.  The exhaust blower capacity is 1200 cfm at
30 inches of water.  If each bag house contains 36
bags four inches in diameter and thirty feet long,
how much dust can be allowed to build up on the
bags before they must be shaken down?  How much
time does this take?

REFERENCES

1.  Briggs, L.W., "Effect of Dust Concentration on
    Cyclone Performance," Trans. Am. Inst. Chem.
    Engrs., Vol. 42, No. 3, pp. 511-526 (1946)
2.  Calvert, S., "Source Control by Liquid Scrubbing,"
    Air Pollution, Ed. by A. Stern, Ch. 46, Academic
    Press (1968)

3. Peters, M.S., "Elementary Chemical Engineering," McGraw-Hill Book Company, Inc., N.Y. (1954)

4. Perry, J.H., "Chemical Engineers' Handbook," 3rd Ed., McGraw-Hill Book Company, Inc., N.Y. (1950)

5. Cornel, D., Knapp, W.G., and Fair, J.R., "Mass Transfer Efficiency - Packed Columns," Chemical Engineering Progress, Vol. 56, No. 7, pp. 68-72 (1960)

6. Morgan, W.D., "The Sorption of Sulfur Dioxide in a Cycled Column: A Comparison Study, " M.S. Thesis, The Pennsylvania State University, CAES Publication No. 116-69 (1970)

7. Robinson, R.G., and A.J. Engel, "An Analysis of Controlled Cycling Mass Transfer Operations," I & E C, Vol. 59, pp. 22-29 (1967)

8. Turk, A., "Source Control by Gas-Solid Adsorption," Air Pollution, Ed. by A. Stern, Ch. 47, Academic Press (1968)

9. Ingebo, R.D., "Drag Coefficients for Droplets and Solid Spheres in Clouds Accelerating in Airstreams," NACA Tech. Note 3762 (1956)

10. Hesketh, H.E., "Cloud Type Atomization of a Liquid Stream by a Gas Stream in a Venturi Scrubber," Ph.D. Dissertation, The Pennsylvania State University (1968)

11. Frederick, E.R., "How Dust Filter Selection Depends on Electrostatics," Chemical Engineering, Vol. 68, No. 13, p. 107 (1961)

12. Iinoya, K., and C. Orr, Jr., "Source Control by Filtration," Air Pollution, Ed. by A. Stern, Ch. 44, Academic Press (1968)

13. Strauss, W., "Industrial Gas Cleaning," Pergamon Press, p. 264 (1966)

14. Ramsdell, R.G., Jr., "Design Criteria for Precipitators for Modern Central Station Power Plants," American Power Conference, Chicago, Ill., (April, 1968)

15. Mednikov, E.P., "Acoustic Coagulation and Precipitation of Aerosols," Translated from the Russion text published in 1963 by USSR Academy of Sciences Press by C.V. Larrick and available through Consultants Bureau, N.Y. (1965)

# CHAPTER XII

## COSTS OF AIR POLLUTION CONTROL

This chapter deals with the direct costs of pollution control which are those expenditures pertaining to control devices and associated facilities necessary to remove the pollution before it enters the atmosphere. Not included are the value of products recovered and the costs of pollution damage (discussed in Chapter V under "Effects"). At this time, costs are usually determined by supply and demand requirements, however, it may be desirable to modify this procedure. For example, utility costs per unit usually decrease as consumption increases. This procedure can encourage large quantity consumers to be wasteful. Reversing this pricing structure to charge more per unit as more of the product is used would encourage conservation.

Technically, pollution can be satisfactorily controlled with the current available knowledge. Practically it becomes a matter of economics, time and the decision to control. The initial investment for pollution control facilities is a heavy economic burden. Some pollution control facilities are economically attractive investments and the pollutants recovered result in a profit, but these facilities are a minority. Much air pollution control is needed, but the extent of this control is highly controversial. This chapter does not discuss philosophical points of air pollution control, but rather will deal with cost as related to direct engineering-type expenditures which must be known in order to estimate preliminary design and operation costs of air pollution control devices.

## 12.1  COSTS

It is usually possible to estimate control costs in a manner similar to that of estimating chemical manufacturing plant costs, because of functional similarity.  Total costs for air pollution control can be divided into two groups-- *capital investment* and *operating costs.  Capital investment* is the total installed cost of a particular facility and includes the following components which are shown as an average percent of the capital investment:  1) major control equipment (35%), 2) auxiliary or accessory equipment (15%), 3) the costs for field installation (20%), 4) project management and engineering (13%), 5) indirect costs such as freight, taxes, contractor overhead, contractor profit, etc. (17%), and 6) an additional 15-20% of total capital costs represent the cost of startup, working capital, capitalized interest and other capitalized costs.  Capital investment or total installed costs of pollution control devices can usually be split equally between the cost of equipment plus auxiliaries (1 and 2 above) which amounts to approximately 50% and the cost of installation (3 through 6) which amounts to the other 50%.  This does not include any solids disposal facilities or site preparation and may or may not include structural supports.

*Operating costs* are the process expenses needed to pay for carrying out the operations as well as for replacing worn out equipment.  These are usually presented on the basis of annual operating costs. For a specific control device such costs vary depending on:  volume of gas cleaned; pressure drop across the system; the total time the device is operated; utility cost for electricity, fuel, water and other raw materials; mechanical efficiencies of motors, fans and pumps; and the gross national product which reflects costs for such items as personnel and services.

Annual operating costs for chemical type operations vary considerably.  In addition to the variables just discussed (which relate to operating cost variations for a single or similar systems depending on input flow variations) operating costs also vary with geographical location because in addition to climatic factors they are a function of such items as local tax structure, financial conditions, interest rates and the existing money market.  Noting all of these variables, the

following is a breakdown of annual operating costs
with *average* values expressed as percent of annual
operating costs:
1) Fixed costs (20%)
   a. Interest, taxes and insurance (about 7% per
      year of capital investment)
   b. Rent
   c. Depreciation (about 15 years for pollution
      control equipment which is approximately
      7% per year of the total installed costs)
   d. Research and development, patents and royalties
      (currently research is <5% of sales)
2) Direct production costs (60%)
   a. Raw materials
   b. Operating, supervising and clerical labor
   c. Maintainence and repair (this amounts to 7%
      of the total capital investment)
   d. Operating supplies and payroll overhead
   e. Power and utilities
      - fuel and power (5% of annual operating costs)
      - steam costs ($0.40/1000 lb steam)
      - electrical energy ($0.02/kilowatt hour
        average; range is 1-3¢)
      - water ($0.15/1000 gallon for drinking
        quality water; range is 10-22¢)
3) General plant overhead (10%)
      - Includes services such as hospitals,
        drafting, safety, recreation, waste
        disposal and control laboratories
4) General administration and office overhead (6%)
      - Includes travel expenses, advertising,
        shipping, and some wages
5) Contingencies (4%)
   Annual operating costs are usually calculated
on the basis of 8,000 operating hours per year and
efficiencies of electrical fans and motors are
assumed to be 60%. Capital investment in manufacturing operations makes up 70% to 90% of the total
investment. The remainder is the working capital
investment which includes such items as raw
materials, supplies, finished and semi-finished
products, accounts receivable, and cash on hand.
   In a normal plant operation, it is necessary
to obtain a percent return on the investment
(percent return is annual profit divided by total
investment times 100). Under normal practices, it
is desirable to have a 10 to 15% annual return on
investment after taxes. As mentioned in the
introduction to this chapter, when pollution
control is considered, it is not always practical

to consider a return on this investment because not all cost factors have been included.

Industrial taxes amount to about 50% of net income. This means that business investments (for pollution control or otherwise) result in a tax savings equivalent to approximately half the expenditure. Similarly, remember that return on investment before taxes is roughly twice the value of the return calculated after taxes.

The Federal and many state governments provide tax relief for industries that install pollution control equipment. Section 169 of the Internal Revenue Tax Reform Act of 1969 permits a fast tax write-off of pollution control facilities. If the facility is certified by both the Environmental Protection Agency and the appropriate state agencies, facilities installed after 1968 can be amortized over a 60 month period. Excepted is any control system that recovers the costs by generating a profit in some manner. Guidelines describing eligible facilities were published in the May 26, 1971, Federal Register.

Thirty-seven states provide tax incentives for pollution control facilities by providing one or more of the following: property tax exemption, sales and use tax exemption, income tax credit, and accelerated amortization. In addition, all but four states (California, Idaho, New Jersey, and Texas) authorize the use of industrial revenue bonds to finance pollution control equipment. These are 15 year bonds which provide tax-free interest to the holders. The advantage of these bonds is they can attract investors, even though their interest rates are about 2% lower than most other bonds, due to their favorable tax free status.

## 12.2  GAS CLEANING COSTS

Table 12.1 is a summary of costs for gas cleaning by various devices. No adsorbers are included in this list. Only particulate matter can be removed with the dry collectors, but wet collectors are capable of removing either particulate or gas or both types of pollutants simultaneously. Dust removal efficiencies are based on a "standard" type dust and listed for comparison only. Obviously, the collector efficiencies will vary depending on the type of dust as well as on dust loading. Collection efficiencies for

gaseous pollutant removal are listed in parentheses
with a (g) behind each number to distinguish from
the particulate removal collection efficiencies.
These approximate values are included only for
comparison. Values in Table 12.1 are based on a
gas flow of 60,000 cfm at 70°F. Costs of similar
collectors vary depending on both 1) the size or
capacity of the unit and 2) the year in which the
unit is purchased and installed. These cost
differences can be estimated by the procedures
outlined in the following sections. Table 12.1
lists the installed costs and includes the basic
device plus auxiliary equipment (such as fans,
pumps and motors)--but does not include solids
disposal equipment and site preparation. Installed
costs are for type 316 stainless construction
materials except as noted and are pessimistically
high, hopefully to reflect maximum expenses. The
costs listed in Table 12.1 are based on 1970
prices using a Marshall and Stevens index of 303.3.
(The Marshall and Stevens index is used to show
relative equipment costs. It is compiled quarterly
for 47 different industries by Marshall and Swift
of Los Angeles.) Inspection of Table 12.1 shows
that the more efficient devices are usually more
expensive to operate.

The total annual cost in dollars per year
given in Table 12.1 includes only the listed
operating costs plus a 10% per year depreciation
cost. This depreciation value is higher than the
7% recommended by the U.S. Treasury Department
Guidelines (4). However, this compensates for
costs not otherwise included. Depreciation rates
will also vary in states providing pollution
control equipment tax write-off advantages and
other incentives. The maintenance charges included
for the filters include one bag change per year
for the envelope type filter and two changes per
year for the shaker and reverse jet filters. Costs
of filter fabrics vary over a wide range depending
on the type, weight and twill of the fabric as
well as the gas flow volume, cloth replacement
frequency, cleaning cycle, time between cleanings
and the operating conditions (temperature and
pressure as well as the specific gas and dust
corrosiveness and abrasiveness).

Packed columns can vary considerably in
installation and operating costs. The data in
Table 12.1 are for counter-current operation of
20 foot high columns. In addition to correcting

TABLE 12.1  INSTALLED AND OPERATING COSTS OF GAS CLEANING BY VARIOUS TYPES OF COLLECTORS (DATA ARE FROM STAIRMAND (1), CALVERT (2), EDMISTEN (3), AND MANUFACTURERS' LITERATURE)

*Conditions:  60,000 cfm gas flow at 70°F*

| Collector Type | Percent Efficiency | Average Pressure Drop, in w.g. | Installed Cost, $ | Power Required, Cost, $/yr. | Clean Water Required, Gal/1000 Cu. Ft. | Water Cost, $/yr. | Maintenance, $/yr. | Total Annual Cost, $/yr. | Total Annual Cost, ¢/(yr Cfm.) |
|---|---|---|---|---|---|---|---|---|---|
| *Dry Particulate* | | | | | | | | | |
| Louver collector | 58.6 | 1.7 | 37,000 | 1,560 | -- | -- | 300 | 5,560 | 9.3 |
| Med. efficiency cyclone | 65.3 | 3.7 | 27,000 | 3,380 | -- | -- | 200 | 6,280 | 10.5 |
| High efficiency cyclone | 84.2 | 4.9 | 52,200 | 4,520 | -- | -- | 200 | 10,040 | 16.7 |
| Multi-cyclone | 93.8 | 4.3 | 56,500 | 3,960 | -- | -- | 200 | 9,810 | 16.4 |
| Electrostatic Precipitator, steel | 99.0 | 0.5 | 181,000 | 2,000 | -- | -- | 1,300 | 21,400 | 36.6 |
| Fabric filter, shaker | 99.7 | 2.5 | 148,000 | 3,740 | -- | -- | 10,000 | 28,540 | 47.6 |
| Fabric filter, envelope | 99.8 | 2.0 | 136,000 | 3,380 | -- | -- | 9,500 | 26,480 | 44.2 |
| Fabric filter, reverse jet | 99.9 | 3.0 | 192,000 | 7,920 | -- | -- | 19,000 | 46,120 | 76.8 |

*Wet Particulate and/or gas (g)*

| | | | | | | | | | |
|---|---|---|---|---|---|---|---|---|---|
| Submerged Nozzle | 93.6 | 6.1 | 71,800 | 5,640 | 0.7 | 2,020 | 700 | 15,540 | 25.9 |
| Impingement Scrubber | 97.9 | 6.1 | 88,600 | 5,800 | 3.6 | 10,380 | 1,000 | 25,840 | 43.1 |
| Wet Dynamic Scrubber | 98.5 | -- | 146,000 | 45,400 | 6.0 | 17,280 | 700 | 77,980 | 130.0 |
| Low Energy Venturi | 99.7(60g) | 20.0 | 83,000 | 20,820 | 5.4 | 15,200 | 1,000 | 45,320 | 75.5 |
| High Energy Venturi | 99.9(70g) | 31.5 | 98,500 | 31,740 | 5.4 | 15,500 | 1,000 | 57,590 | 96.0 |
| Gravitation spray tower | 96.3(55g) | 1.4 | 68,800 | 8,480 | 18.0 | 51,200 | 1,000 | 67,560 | 113.0 |
| Packed column,* stainless packing | 70(g) | 10.0 | 240,000 | 12,520 | 8.0 | 22,200 | 800 | 59,520 | 99.2 |
| Packed column,* steel w/ceramic packing | 70(g) | 10.0 | 138,000 | 12,520 | 8.0 | 22,200 | 800 | 49,320 | 82.2 |
| Packed column,* steel w/PVC packing | 70(g) | 10.0 | 62,400 | 12,520 | 8.0 | 22,200 | 600 | 41,560 | 69.3 |
| Bubble cap column,* stainless | 60-80(g) | 16.0 | 108,000 | 16,400 | 8.0 | 22,200 | 450 | 49,850 | 83.1 |
| Sieve plate column | 60-80(g) | 8.0 | 91,000 | 8,110 | 8.0 | 22,200 | 450 | 39,860 | 66.5 |

*Conditions: 60,000 cfm gas flow at 70°F.*

*Plate columns are for 10 plates and packed columns are for 20 ft of height

for size according to Section 12.3.1, columns under 6 feet in height may be more economically operated as cross-flow systems (5). Plastic packing, when it can be used, is virtually unbreakable plus light in weight, making it possible to use thinner column shells and thereby gain added cost savings. Fuel operating costs mentioned under annual operating costs usually apply to steam and heating requirements. Edmisten (3) shows that the fuel costs for afterburners range from 0.014¢/cfm per hour to 0.057¢/cfm per hour for systems with and without (respectively) heat exchangers using 50% excess air, if there is no heating value in the pollutants.

Table 12.1 assumes a 60% fan or motor efficiency is for all systems. Devices such as Venturi scrubbers, which require higher pressures, frequently have added costs due to greater pumping losses. For example, in a Venturi scrubber at 30 inches of water, it is estimated (2) that 110 kilowatt hours of energy are required per million cubic feet of gas cleaned. Water pumping horsepower at 45% efficiency is approximately equal to $(1.25 \times 10^{-4})(\Delta p)(q)$ where $\Delta p$ is the pressure drop of the water in psig and $q$ is the water flow rate in gpm. The horsepower requirement for moving the gas stream at 60% efficiency is approximately equal to $(3 \times 10^{-4})(\Delta P)(Q)$ where $\Delta P$ is the pressure drop of the gas in inches of water and Q is the volumetric gas flow rate in cfm. The water costs in Table 12.1 are based on drinking quality water, so any absorbed product would not be contaminated with impurities. Note that if the Venturi scrubbers or the gravitational spray tower are operated on recirculated water, the annual water costs could be 25% of the costs shown.

The costs of simple settling chambers average approximately $18 per cubic foot of capacity for the installed devices not including supports, ducts, fans or any other accessories. Usually, these installations are part of a system that already requires a fan for operation (some existing systems have no spare capacity and settling chambers cannot be added after the initial design without changing the blowers). No new facility should be designed without cleaning devices as part of the initial installation, thereby avoiding any significant cost increase that might be incurred by replacing the blower.

Costs of pollution control expressed as percentage of profit reduction vary enormously from industry to industry. Economic studies made in Pennsylvania in 1968 by the author indicated a cost from 0.8 to 1.5% to control the pollution releases from medium to large size chemical and steel industries. On the other hand, it would cost as much as 10.4% to control pollution from smaller (but similar) industries. This variation indicates that pollution control cost assistance to certain smaller commercial operations may be necessary to prevent financial ruin.

## 12.3 COST DATA EXTRAPOLATION

Installed costs of air pollution control equipment can be estimated for various sizes of equipment using an exponential type cost relationship. These installed costs can then be corrected for current economic conditions. Utility costs such as power, water and fuel, must be accounted for by calculating the amount of energy required (e.g., horsepower, quantity of make-up water, amounts of fuel, etc..) and then estimating the costs of these utilities using the values available from Sections 12.1 and 12.2.

### 12.3.1 EQUIPMENT SIZE AND MATERIAL OF CONSTRUCTION

It is possible to estimate the costs of various sizes of a specific type of equipment by several procedures. If the costs of two or more different size units are known, the log of the cost versus the equipment capacity can be plotted as a straight line. In the absence of this information, it is possible to estimate approximate costs using an exponential (logarithmic) relationship such as:

$$I_b = I_a \left(\frac{C_b}{C_a}\right)^n \qquad (12.1)$$

where: $C$ = capacity or size of equipment a or b
$I$ = capital investment of equipment
$n$ = sizing factor exponent

This exponent based technique provides a fast yet reliable enough procedure for making preliminary cost evaluations.

In the absence of other data, it is frequently useful to use the Williams' six-tenths rule which suggests that a common value of n equal to 0.6 is often a suitable scaling (sizing) factor. Industrially, scale up factors for specific devices can be obtained from cost data. An example of this is the chemical company data listed in Table 12.2 (6,7). Note that the value of the sizing factor exponent (n) can vary considerably from the value of 0.6 suggested by Williams, however, 0.6 is a good average figure. Discretion should be employed in using the 0.6 exponent and estimates beyond a 10-fold range of capacity are subject to considerable doubt.

TABLE 12.2   SIZING FACTOR EXPONENTS (n) FOR SOME AIR
             POLLUTION CONTROL DEVICES (6,7)

| *Device* | *Material of Construction* | *Physical Description* | | *n* |
|---|---|---|---|---|
| Sieve plate tower | Carbon steel | 10-20 plates | 60-120" dia. | 1.40 |
| "        "      " | "            | 21-45 | 30-60 | 0.28 |
| "        "      " | "            | 21-45 | 60-200 | 0.71 |
| "        "      " | "            | 46-75 | 24-60 | 0.30 |
| "        "      " | "            | 46-75 | 60-200 | 0.62 |
| "        "      " | 304 SS       | 10-35 | 30-120 | 0.65 |
| "        "      " | "            | 36-75 | 30-120 | 0.38 |
| "        "      " | 316 SS       | 10-80 | 18-60 | 0.34 |
| "        "      " | "            | 10-80 | 60-200 | 0.72 |
| Bubble cap tower | Carbon Steel | - | 36-100 | 0.86 |
| "      "      " | Stainless (SS) | - | 36-100 | 0.52 |
| Packed tower | Stainless | - | 5-36 | 0.35 |
| "      " | "          | - | 36-100 | 0.85 |
| Column (Tower) Shells | 316 SS | 5-20 ft high | | 0.75 |
| Column Packings | Porcelain | Raschig Rings | | 1.3 |
| "       "       | "         | Intalox Saddles | | 1.2 |
| "       "       | Polypropylene | Pall Rings | | 1.4 |
| "       "       | "         | Intalox Saddles | | 0.76 |
| "       "       | 316 SS    | Pall Rings | | 1.7 |
| "       "       | "         | Propalc or Hyperfil | | 1.5 |
| Cyclone | Steel | Dust Collector or Scrubber | | 0.72 |
| Cyclone | 316 SS | Scrubber | | 0.85 |
| Venturi | Steel or 316 SS | Scrubber | | 0.73 |
| Impingement Plate | " | Scrubber | | 0.73 |

Calvert (2) suggests some of the following relative cost conversion factors to account for various materials of construction: carbon steel (basis = 1.0), copper (1.4), polyvinyl steel (PVS) (1.4), aluminum (1.5), lead (1.6), 304 stainless (2.3), 316 stainless (2.7), 70-30 copper-nickel (2.7), 317 stainless (3.4), monel or nickel (3.0), hastelloy (3.5) and monel 400 (4.8).

## 12.3.2 COST INDEX

Costs of purchasing equipment and costs of installing facilities vary with the ever changing economic conditions. To approximate current costs, it is necessary to multiply the old equipment cost value by the ratio of some current index value to the index value applicable during the year the old cost was obtained:

$$P_b = P_a \frac{N_b}{N_a} \qquad (12.2)$$

where: P = cost of equipment purchased
N = cost index values for year a or b

Although indices have always been severely criticized, they represent the only means to achieve estimates particularly of a primary nature within a reasonable length of time and without undue effort. Some of the numerous indices available are: The Marshall and Stevens All-Industry and Process-Industry Indices, the Nelson Refinery Index, the Engineering News Record Construction Index and the Chemical Engineering Plant Cost Index. The Marshall & Stevens (M&S) Equipment Cost Index for the average of eight process industries is perhaps the most readily available. This index was started in 1926 at a base value of 100 and is compiled quarterly by Marshall & Swift. It is published monthly by such publications as *Chemical Engineering*. Figure 12.1 shows the M&S Equipment Cost Index for the average of eight process industries. It includes the 1971 average purchase cost index value of 321.3 and extrapolated values beyond 1971. These values can be used for N in Equation 12.2.

It is possible to extrapolate the total capital costs using the economic indicators given, remembering that equipment costs approximately equal installation costs and assuming equivalent changes for both.

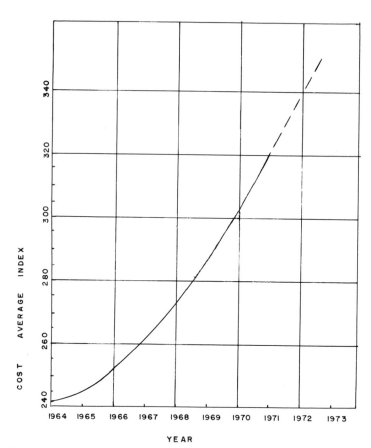

Figure 12.1    Marshall and Stevens Process Industries
Average Equipment Cost Index (1926 = 100)

The relative consumer price index has also been
increasing at about the same rate as the Marshall
& Stevens Index; however, the gross national
product which was increasing at a rate of from 6-9%
per year up to 1966 has not been increasing rapidly
recently.  It is estimated that there will be an
annual GNP increase of from 2-6% up to 1976.  If
this happens, the actual Marshall & Stevens Index
Values beyond 1971 may be less than the indicated
extrapolated values shown in Figure 12.1.

The "time value" of money has not been included
in these discussions. The reader who wishes to
consider this important factor is referred to texts
such as that by Taylor (8). Specific pollution
control costs for various industries and geographical
locations have also been compiled (9).

QUESTIONS FOR DISCUSSION

<u>1</u>. Discuss the relative advantages and disadvantages
of the various cost indices.
<u>2</u>. What is the most accurate way of obtaining an
estimate of purchased, installed and operating costs
for air pollution control equipment?
<u>3</u>. How would you account for the cost factors not
included, such as sale of recovered pollutants,
prevention of soiling and other effects?
<u>4</u>. How does dust loading enter into the annual
operating costs?
<u>5</u>. On the average, costs increase as the equipment
size increases. However, the reverse of this is
true for some small devices. Explain why.
<u>6</u>. What happens to operating costs per quantity of
thruput material as the amount of material handled
increases? Why?

PROBLEMS

<u>12.1</u> Estimate the cost of installing a carbon steel
Venturi scrubber for handling 30,000 cfm of gas
in 1971.
<u>12.2</u> Estimate the annual operating cost of the
Venturi scrubber in Problem 12.1.
<u>12.3</u> What is the most economical way of removing
98% by weight of the dust from 250,000 cfm gas
stream if the dust load in the gas stream is light?
The dust has a mean diameter of 0.6 microns and a
standard deviation of 1.3. State all assumptions.
<u>12.4</u> Estimate the cost of a packed column with
304 stainless packing to handle 25,000 cfm of gas.
It is estimated that 12 transfer units will be
required for the particular wet scrubber in question
<u>12.5</u> Estimate the cost of the dust collector for
Problem 11.4 if it is made of PVS.
<u>12.6</u> How many months will it take to complete the
construction of the column in Problem 12.4? For
projects up to $10 million, the normal completion
date may be estimated using:

$$m = 13 + 8 \log C \qquad (12.3)$$

where:   m = months from date of cost estimate
         C = project cost in millions of dollars
12.7  Estimate the cost of the scrubbing tower
used in Problem 11.9 (the properly sized unit) if
the material of construction is steel.

REFERENCES

1.   Stairmand, C.J., "The Design & Performance of
     Modern Gas Cleaning Equipment," paper read
     before Institute of Chemical Engineers, London
     (1955)
2.   Calvert, S., "Source Control by Liquid Scrubbing,"
     Air Pollution, Ch. 46, edited by A. Stern,
     Academic Press (1968)
3.   Edmisten, N.G., and F.L. Bunyard, "A Systematic
     Procedure for Determining the Cost of Controlling
     Particulate Emissions from Industrial Sources,"
     JAPCA, Vol. 20, No. 7, pp. 446-452 (1970)
4.   "Depreciation Guidelines & Rules," U.S. Treasury
     Department, Internal Revenue Service Publication
     No. 456, Wash., D.C. (1962)
5.   Hanf, E.B., "A Guide to Scrubber Selection,"
     Environmental Science & Technology, Vol. 4, No.
     2, p. 110-116 (1970)
6.   Chase, J.D., "Plant Cost vs Capacity:  New Way
     to Use Exponents," Chemical Engineering, Vol.
     77, No. 7, pp. 113-118 (1970)
7.   Drew, J.W. and A.F. Ginder, "How to Estimate the
     Cost of Pilot-Plant Equipment," Chemical
     Engineering, Vol. 77, No. 3, pp. 100-110 (1970)
8.   Taylor, G.A., "Managerial and Engineering
     Economy", D. Van Nostrand Co., Inc. (1968)
9.   Fogel, M.E., et. al., "Comprehensive Economic
     Cost Study of Air Pollution Control Costs for
     Selected Industries and Selected Regions, NAPCA,
     USPHS, Dept. of HEW, PB 191 054 (1970)

# CHAPTER XIII
## SAMPLING AND ANALYSIS

The National Air Pollution Control
Administration (NAPCA) estimated that as of June,
1970, there were 14,685 air quality sample
collecting and monitoring devices available. This
includes the static devices, of which there are
3,147 for gases and 4,437 for particulates, plus
mechanized and automatic devices. The mechanized
devices are powered (usually by electricity) and
can mechanically accumulate samples either
consecutively or intermittently for subsequent
laboratory analyses. Mechanized devices include
2,387 gas type and 2,958 particulate type devices.
Automatic devices continuously sample and analyze
the air to produce numerical or visual results for
direct evaluation. There are 1,092 gaseous devices
and 368 particulate devices of this type. No
attempt is made in this chapter to discuss individual
analytical systems.

It is predicted that the demands for source
analyses instruments will continue to increase
through 1980 but that the demand for ambient
analyses devices will peak in 1972. The demand
for instruments specifically intended for auto
emission analyses should reach a maximum in 1975.
Even though a large number of analytical devices
are available, we still lack instruments to perform
certain specific analyses.

It is important to recognize that no matter
what analytical instrument or procedure is used, if
the sample is not obtained meaningfully, even the
most accurate analysis has no significance. It is
the intent of this chapter to discuss procedures by
which meaningful samples may be obtained and then to
mention briefly some general types of analytical
devices. Several simple laboratory experiments are
included in the problem section of this chapter.

## 13.1   AIR QUALITY CYCLES

The concentration level of pollution in the atmosphere is a function of both place and time. The place or geographical factors are mainly: 1)   location of the receptor relative to the pollution source--the atmosphere dilutes the pollution so that beyond the maximum ground level concentration, the pollution level decreases with increased distance from the source; rural areas can usually be expected to have cleaner air if they have fewer pollution sources; and locations downwind from metropolitan regions can be expected to have higher concentrations than those located upwind; and 2)   type of terrain--mountain ranges, valleys, shorelines and other locations can produce higher or lower levels of pollution as discussed in Chapter III.

Pollution concentrations may vary in a cyclic manner.   These cycles are not always as obvious as the geographical variations, but nevertheless can exist.   Cyclic variations can be attributed to: 1)   source variations, 2)   meteorological variables, 3)   atmospheric removal mechanisms, and 4)   human or mechanical variations.   These variations are explained in the following paragraphs.

It is possible for one or more (even all) of the four factors which cause air quality cycles to act together to produce a specific type of pollution cycle.   A source variation would result, for example, when a stone quarrying operation begins operation in the morning and discontinues operation at the end of the day.   Another source variation example occurs when a power generating facility has more demand for power in the evening than any other time of the day.   In addition to the power demand changes, meteorological factors would probably cause higher concentrations in the evening and again during the early morning hours, due to presence of an inversion (or at least stable atmospheric conditions).   This could result in a daily cycle such as shown in Figure 13.1 where there are two periods of the day which have high concentrations and two periods that have low concentrations.

Another example of a daily air quality cycle that includes source variations, meteorological variables and atmospheric removal mechanisms is presented in Figure 13.2.   Source variables, in this example, can occur when automobiles are used in the morning and evening hours to drive to and from work.   As a result, the concentration of nitric oxide builds up during these times. As the sun converts the nitric oxide to nitrogen dioxide, the concentration increases until the photochemically catalyzed reactions decrease the amount of $NO_2$ present and create oxidants as shown in the diurnal cycle of Figure 13.2. It is obvious that samples taken at 2:00 in the afternoon will not reflect an average concentration of nitrogen dioxide, nor will samples at that same time reflect the average daily concentration of oxidants. These source variables also combine with location factors so that the maximum concentration of the photochemical smog produced in this example occurs several hours after the emissions are released (i.e., occurs at mid-day) at a location that is miles downwind.

It should be easy to recognize that there are annual air quality cycles. For example, the sulfur dioxide concentration is highest during the winter time when more sulfur bearing fuels are burned for heating purposes. In contrast, the oxidants have the highest concentration during the summer when

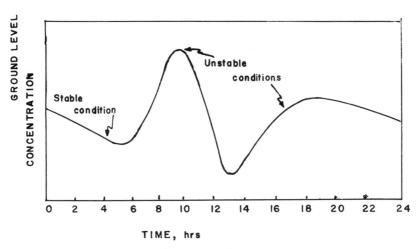

Figure 13.1   Daily Air Quality Cycle from Source and Meteorological Variables

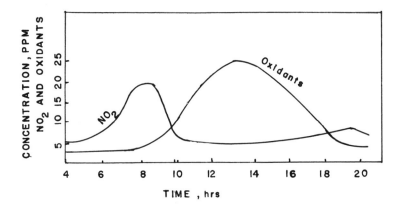

Figure 13.2   Daily Air Quality Cycle from Source
and Meteorological Variables and
Atmospheric Removal Mechanisms

there is more sunlight for photochemically
catalyzed reactions.  This suggests that annual
pollution concentration values can be averages of
greatly varied ambient concentrations.

Data can be obtained to show air quality cycles
that cannot be explained by any of the first three
time factors.  When this is the case, it is
worthwhile to determine if the cycle is a result
of human or mechanical variations.  For example,
different operators making analytical measurements
can consistently obtain significantly different
results.  Should these operators work on a "shift"
basis, there would be a noticeable change in the
data for every shift.  Other factors, such as daily
voltage drops or temperature and humidity changes,
can account for cyclic variations in the sampled
data.  Emission heights are also responsible for
some air quality cycles, even though all other
factors are considered.  For example, low stacks
result in higher ground level concentration and
high stacks, on the other hand, should provide a
lower concentration at a given distance from the
stack and for a given quantity of emissions.
However, during the morning hours when the sun
rises, fumigation can occur which temporarily

results in extremely high concentrations from high
level emissions. Sampling at the time and location
where fumigation occurs would not give representative
daily averages.

Periods of inversions produce abnormal
conditions. The number and frequency of inversions
will be greater during the fall of the year than at
other times as discussed in Chapter III.
Interestingly, recent data obtained using the
scanning electron microscope show that particle
size distribution and mean diameter are not a
function of time of day and therefore do not exhibit
cyclic changes (1).

The reader should be aware by now that samples
taken once or twice a day certainly may not represent
daily averages. On the other hand, it may not be
practical to sample continuously, so some type of
a time interval and averaging procedure must be
established. Meteorologists refer to the three-
minute wind cycle which primarily mixes the material
in the atmosphere. The one hour winds are the winds
that actually carry the material from the source to
the sampling site. As a result, one hour sample
intervals are frequently used for atmospheric
analysis. The one hour samples are then averaged
using arithmetic or geometric averaging procedures
to obtain daily averages. In turn, the daily
averages are then averaged in the same way to
produce the annual means.

Plots of pollution concentration versus the
length of the sampling time can be made on log-
probability paper for a given sampling location.
A long sampling time means that the pollution sample
is composited over a long period, then analyzed as
an average pollution concentration. As duration
of sampling time increases, the standard geometric
deviation ($\sigma$) decreases resulting in variations of
the geometric mean. For example, a typical value
of a 5 minute $\sigma$ is 2.0 and a 1 day $\sigma$ is 1.5. The
arithmetic mean concentration of course does not
change with length of sampling time. The geometric
mean ($G_M$) and arithmetic mean ($A_M$) concentrations
are related by the expression:

$$\ln A_M = \ln G_M + \tfrac{1}{2} (\ln \sigma)^2. \qquad (13.1)$$

13.2 SOURCE SAMPLING

It is necessary to obtain good source data to
determine whether or not air pollution emission

standards are being exceeded.  Source sampling is
essentially stack sampling and for the most part,
consists of the following good engineering
procedures.  For example, near sampling sections,
there should be no eddy formation resulting from
turbulence caused by bends or obstructions in the
system.  It is recommended that sampling points
should be at least 15 pipe diameters downstream
from any bend or obstruction and there should be
no bends or obstructions for at least 10 pipe
diameters downstream from the sampling point.  In
addition, the sampling device should result in as
little disturbance to the system as possible and
must be aimed directly into the stream.

The concentration of material in a stack is not
uniform throughout the cross section of the stack
therefore, it is necessary to make a traverse of the
stack and average the resulting values.  The least
number of samples that should be taken from a stack
is 16, with a greater number of samples being
required from "large" stacks.  One way to traverse
a stack is to divide the diameter of the stack into
eight points representing four spaces of equal area.
This can be done for a circular stack by making the
first point 0.033 diameters distance from the *inside*
of the wall, the second point 0.105 diameters, the
third 0.194 diameters and the fourth 0.323 diameters.
The remaining four points are made the same distances
from the other side (or measuring from the same side
the distances are 0.677, 0.806, 0.895 and 0.967
diameters).  Samples are taken at these eight points
and then eight more samples are taken at the same
distances from the wall but perpendicular to the
line (in the same plane) from which the first
samples were taken.  Data taken at these 16 points
can be considered of equal value because each sample
is taken from the center of the same size cross
section of area.  It should be noted that when this
sampling technique is used for velocity measurements,
for example, by using pitot tube measurements, the
velocity is determined using the square root of the
pitot tube pressure differential.  In this case, it
is necessary to average the pitot readings by
*averaging* the *square roots* of all values to obtain
the *average square root*.  When rectangular ducts
are sampled, simply divide the entire area into
about 12 equal area sampling squares and average
the resultant data.

It has been mentioned that the sampling device
should not disturb the system.  This can only be

accomplished by the use of isokinetic sampling even
though the smallest, most streamlined sampling probe
is used.  Isokinetic sampling is especially important
for particulate sampling and for sampling dense
fluid streams.  Isokinetic sampling is defined by
the name itself, which means maintaining the same
kinetic velocity inside the sampling device as in
the stream being sampled.  At 100% isokinetic
sampling, the concentrations of materials in the
gas stream and in the sample stream are the same.
When the velocity in the sampling device is less than
the velocity in the gas stream, the sampling rate
is less than 100% isokinetic.  This causes the
smaller particles to be deflected around the sampling
head.  The sampling analysis will then be in error
because there is an excess of large particles (the
sample mean diameter will be large and standard
deviation will be low).  If too high a sampling
velocity is used, the sampling rate will be greater
than 100% isokinetic and an excessive number of
smaller particles will be drawn into the sampling
device.  The sample analysis will then erroneously
show too small a mean diameter and a high standard
deviation.

Instead of devising procedures for correcting
data obtained at non-isokinetic conditions, it is
recommended that *isokinetic sampling* be performed.
To accomplish this, it is necessary to have a flow
metering device located next to the sample nozzle.
This is most conveniently done using a pitot tube
and a sample nozzle, or a commercial device known
as a pitobe which is shown in Figure 13.3.  The
particulate sample is usually obtained by impingers
or some other device before the gas flow from the
sampling nozzle is measured.  This means there may
be a decrease in temperature of the sampled gas with
a corresponding decrease in the sampling volumetric
flow rate.  It is necessary to measure the stack
temperature as well as the temperature at which
the sample gas flow is measured so that the sample
flow can be corrected to stack conditions to be
sure of having the proper gas velocity in the
sampling nozzle.  These calculations can be made
using standard orifice and pitot tube equations or
by using any other accurate flow metering devices.

Stack velocities will be different at the
different sample points and, in addition, can
change because of flow variations in the system
being sampled.  This means that considerable time
can be wasted completing calculations to establish

isokinetic conditions.  A table or a nomograph for
the sampling nozzle should be prepared in advance
to facilitate the sampling.  The pitobe shown in
Figure 13.3 has three different nozzles which can
be used to help obtain the desired sampling velocity.
This commercial sampling device is electrically
heated and insulated to help reduce temperature
drops and to prevent condensation in the sampling
tube.  The system can handle flows from 1,000 to
10,000 ft/min at temperatures up to 1200°F.
    Sampling times of up to *two minutes* should be
used at each sampling point.  If the dust loading is
not excessive, composite samples can be taken across
the stack to facilitate analysis procedures *if* the
sampling time at each point is made equal.
    When isokinetic sampling cannot be used, it is
possible to estimate the isokinetic percentage
using (2):

$$\% \text{ isokinetic } = (100)(M_a/M_c) \qquad (13.2)$$

where:  $M_a$ = mass emission rate of pollutant
             on an area basis
        $M_c$ = mass emission rate of pollutant
             on a concentration basis

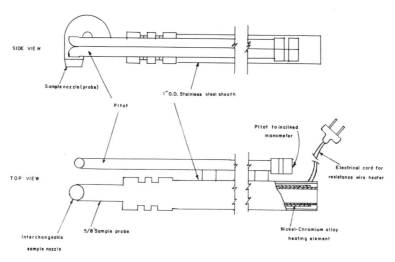

Figure 13.3   Commercial Pitobe for Isokinetic
              Sampling

The mass emission rates are determined by:

$$M_a = \frac{M}{t}\left(\frac{A_s}{A_n}\right) \qquad (13.3)$$

$$M_c = M\,A_s\left(\frac{V_s}{V_n}\right) \qquad (13.4)$$

where: $M$ = mass of pollutant collected in sampling time $t$
$A_s$ = area of stack cross section
$A_n$ = area of sam;ling nozzle
$V_s$ = stack gas velocity
$V_n$ = sample gas volume converted to stack conditions

The manual "Source Testing for Air Pollution Control," by Cooper and Rossano, Environmental Science Services Div., Wilton, Conn. (1971) is recommended as a detailed reference on this subject.

## 13.3   SAMPLING LOSSES

In addition to the normal considerations of keeping sampling lines short to minimize sampling time lags and making certain that the sampling devices, lines and measuring instruments are clean, other considerations can contribute to sampling errors. These include chemical reactions, absorption and adsorption. When heated sampling tubes and lines are used, it is necessary to prevent localized over-heating which can cause not only polymerization of organic materials, but can also increase reaction rates between any pollutants coexisting in the sample. Adsorption, including chemical adsorption, causes losses. For example, tests run on sulfur dioxide showed that much of the $SO_2$ in the sample was lost due to the initial adsorption on tubing made of polyvinyl chloride, polyethylene and tygon. Tubing of aluminum and copper should not be used at all because of the adsorption and resulting chemical reaction with the tubing which creates permanent, continuous sample losses. Teflon and glass tubing can be used for sampling $SO_2$ without loss. The use of polyvinyl chloride tubing is satisfactory if the tubing is conditioned for a one hour period using gas containing at least 0.1 ppm $SO_2$. Polyethylene and tygon coult be used; however, their conditioning periods amount to approximately 15 hours at 0.2 ppm.

Copper should obviously not be used in ammonia samples because of the chemical reaction and formation of copper ammonia complexes; aluminum tubing should not be used when oxidants, acids or bases are present; glass tubing should not be used with fluorides. Compatibility checks should always be made when specifying type of tubing and other materials in the sampling system. It should be noted that stainless steel may not be satisfactory for sulfur dioxide or sulfuric acid type samples and glass fiber filters are known to adsorb sulfur dioxide. For short sampling times and peak $SO_2$ concentrations, a Tenite plastic holder with a Nuclepore membrane filter is recommended (3). Teflon and stainless steel can be used in ambient $SO_2$ sampling where the concentrations of $SO_2$ are low, but both of these materials are relatively expensive.

It is not necessary to eliminate adsorption, however, it must be recognized and the sampling loss must be accounted for while the system is being conditioned (saturated). A further complication arises from the fact that if the sample concentration fluctuates, the adsorbed material may be released at times of low concentration, giving a higher apparent concentration as a result of this desorption.

## 13.4  EQUIPMENT

The fantastically high number of devices available for analyzing ambient air was pointed out in the introduction of this chapter. In addition, it is necessary to have other sampling facilities and accessories. Information concerning suppliers of these various items can be obtained from publications such as: "The Annual Air Pollution Control Association (APCA) Product Guide" published by APCA, 4400 Fifth Avenue, Pittsburgh, Pa. 15213; "The Annual Pollution Control Directory" contained in Environmental Science and Technology published by Reinholdt Publishing Corporation (through the American Chemical Society); "The Annual Environmental Engineering Directory" contained in Chemical Engineering published by McGraw-Hill Publications; and the "Annual Contamination Control Directory" published by Contamination Control magazine. In addition, Chemical Week contains an annual buyers' guide of chemicals, equipment and packaging; Reinhold Corporation publishes the annual Chemical Engineering Catalog of chemical materials and process equipment; and Industrial Research, Inc. publishes

the "Industrial Research Buyers' Guide Yearbook."
In addition, many pollution control journals include
air pollution advertising literature.
    Certain *basic* items necessary for air pollution
sampling are listed as follows:
1.  Vacuum pumps (oilless and/or leak free types)
    for drawing the sample into analysis devices.
2.  Small flow meters capable of handling up to
    10 liters per minute (20 ft³/hr) gas flow
    and devices to calibrate these flow meters,
    such as calibrated nozzles, calibrated
    rotometer kits, wet test gas meters or
    limiting orifices (note that glass capillaries
    can be useful limiting orifices when small
    flows are required).
3.  Pitot tube and inclined manometer for gas
    velocity measurements.
4.  Drying tubes, connectors (T's, L's and
    quick-disconnect joints) and tubing.
5.  Filter holders and filter papers for both
    inline filtering and for obtaining dust
    soiling measurements of particulate matter
    in the ambient air.
6.  Particulate size distribution measuring
    devices such as impingers or inertial
    impactors.
7.  A microscope for analyzing particulate
    samples. A valuable companion to the
    microscope is The Atlas by W. C. McCrone (4),
    which contains photomicrographs for
    particle identification.
8.  Gas sampling tubes for collecting gas samples.
9.  Analytical balance.

In addition, thermometers, glassware, stopwatch and
other standard supplies should be available.
Sampling instruments can be added to include specific
sampling requirements. It would be worthwhile to
have access to a colorimeter and a method of
determining pH if wet chemical analyses are planned.
Distilled or deionized water is a definite must for
analytical work of almost any type.


13.5  PROCEDURES

    Before starting to take either ambient or stack
type measurements, valuable time can be saved if
thorough advance preparation is made. This includes
not only considering the factors already mentioned
in this chapter, but suggests consideration be given

to obtaining atmospheric information and emission data.  It is possible to estimate atmospheric variables such as wind speed, wind direction, mixing height (approximately equal to the top of the lower cloud layer) and inversion conditions. These data may be accurately obtained from *nearby* airports, or other measuring stations.

It is valuable to know in advance approximately how much and what type emissions are likely to come from certain operations.  This information can be checked beforehand with emission guides and other publications frequently referred to in this text.  Other sources which could be valuable in supplying this information include chemical or chemical engineering publications such as References (5) and (6).

It is frequently not possible to obtain meaningful analyses using stored air pollution samples.  This means that most analyses should be made directly at the sampling site.  Procedures for instrumental measurements should be supplied, with the devices.  In addition to instrumental analysis methods, wet chemical analyses can also be used. In fact, most instruments are calibrated using wet chemical procedures.  The selected methods of measurements by wet chemistry for six classes of gaseous pollutants as well as for the measurements of sulfate and nitrate on particulates are described in Reference 7.  This reference discusses the West and Gaeke and the hydrogen peroxide methods for the determination of sulfur dioxide, as well as the Saltzman method for determination of nitrogen dioxide and nitric oxide.  In addition, methods of analyzing for oxidants, aliphatic aldehydes, acrolein and formaldehyde are included.  The Federal Register and state standards list procedures which can be or are required to be used for specific sources and ambient analyses.

Other references specifically devoted to wet chemistry procedures for analyzing air pollutants are Appendix B of Reference 8 which discusses analysis of both sulfur dioxide and sulfuric acid mists, and Appendix A of Reference 9 which discusses the sampling and analytical techniques for measuring nitric oxides as generally used in the nitric acid manufacturing industry.  Special grades of chemicals are sometimes required to carry out the various wet chemistry analyses.

A procedure for determining the soiling index of particulate matter in the air is to optically

determine the surface area of a composite sample.
The results are then given in units of coefficient
of haze (Coh). Air samples are drawn through a
filter paper of known cross section area at a
controlled velocity for a predetermined period of
time. This data is used to convert the volumetric
flow of air into columns of air the diameter of the
filter spot. The number of 1,000 ft. long columns
passing through the filter are then used to report
the soiling density as Cohs/1,000 lineal ft. Numerical value of the Coh is determined by:

$$Coh = \frac{O.D.}{0.01} \qquad (13.5)$$

The optical density (O.D.) is determined using 4,000
A   frequency light by:

$$O.D. = \frac{intensity\ of\ light\ through\ clean\ paper}{intensity\ of\ light\ through\ dirty\ sample}$$

$$= \log \frac{100}{\%\ transmission} \qquad (13.6)$$

Combining both of these equations and expressing the
results per 1000 lineal feet of air gives:

$$Coh/1000\ ft = \frac{(area\ of\ soiled\ spot)(100,000)}{(cfm\ of\ air\ sampled)} \log \frac{100}{\%T}$$

$$(13.7)$$

Analytical facilities for this must include a photo-
electric cell for indicating the intensity of light
passing through the clean paper, and the intensity
of the light passing through the section of filter
paper with the captured dirt.

Odors are difficult pollutants to quantitatively
analyze because they exist at such low concentrations.
One of the more sensitive and less complicated
devices for odor analysis is gas chromotography
using hydrogen flame ionization detection. This
method is applicable for all organic gases and is
capable of measuring down to $3 \times 10^{-12}$ g/sec
(equivalent to tenths of a ppb reported as propane).
Argon ionization gas chromotography is 10 times more
sensitive and is applicable for most organic and
inorganic gases, however, this procedure is more
complicated. Odor analysis devices include the
crude hot wire halogen leak detectors as well as
flame photometric, thermal conductivity,

electrolytic conductivity, and electron capture detection. Special instruments are obviously required for adequate odor analysis.

In general, some of the instrumental analysis procedures used for measuring gaseous pollutants are:

1. Conductometric analysis measures the electrical conductivity of a solution--an example is the oxidation of $SO_2$ to sulfuric acid which is a good electrical conductor.

2. Coulombometric analysis is the measurement of the quantity of current flowing--an example of this is the ozone monitor which oxidizes potassium iodide solution with ozone and then regenerates the potassium iodide at a platinum electrode, where the current intensity required for regeneration is calibrated in terms of concentration of ozone (carbon anodes complete the redox reaction forming carbon monoxide).

3. Infrared absorption measures the amount of pollution present as a function of the infrared (long wave) light absorbed by the pollutants-- the resonance frequency of an electrical tank circuit which varies depending on the capacitance changes incurred when infrared light intensity changes are used to indicate the amount of pollution present. The light absorption depends mainly on the covalent bonds of the pollutants present.

4. Flame ionization measures the amount of electrons given off by ionized pollutants and presents this information as the quantity of pollution present. An example of this is the hydrocarbon analyzer that burns hydrocarbons in a nitrogen-hydrogen flame to produce ions which, when collected at the anode, take up electrons that constitute the current flow.

5. Flame photometry measures the spectral intensity and frequency of the light given off when samples are sprayed into a flame.

6. Colorimetry measures the absorption of 5600 Angstrom (short wave) or other wavelength light by various pollutants and concentrations of pollutants. The intensity of the unabsorbed light is then compared with a previously determined calibration curve.

Mass spectrometry and flame ionization emission spectroscopy, and gas chromatography as mentioned above, are also frequently used techniques for analyzing pollutants.

Analysis of air pollutants is becoming more elaborate with the use of new techniques and the combining of procedures. Lasers, for example, are being used both for particulate *and* gas analyses. The scattering of light due to the presence of particles indicates size and even size distribution of the particulate matter. It has recently been determined that laser light scattering due to the presence of particulate matter is related to the volume of aerosol present per volume of air (10). Section 7.1.1 lists other methods used to size particulates.

Laser pulses directed at gas mixtures cause the molecules to re-emit photons of different wave lengths. A spectrum analysis of the re-emitted light is used to "fingerprint" the particular types of gases present. Attempts are now underway to conduct experiments on air pollution measurements from satellites orbiting in space. One of the most complete analytical devices available is the molecular rotational resonance (MRR) spectrometer which theoretically has the capability of analyzing for any substance that exists. This device can make both qualitative and quantitative measurements.

QUESTIONS FOR DISCUSSION

<u>1</u>. What type of air quality cycles exist?
<u>2</u>. What causes each of the cycles in No. 1?
<u>3</u>. What considerations should be kept in mind when preparing to make ambient air analyses?
<u>4</u>. What considerations should be kept in mind when preparing to make stack analyses?
<u>5</u>. Define optical density, transmittance and absorbance.
<u>6</u>. How would you determine the number of Cohs for a 1-inch diameter sampling tape?
<u>7</u>. Name as many methods as you can for qualitative and quantitative analyses of: a) $SO_2$  b) $NO_x$ c) sulfuric acid  d) carbon monoxide  e) oxidants f) hydrocarbons  g) hydrogen sulfide  h) ammonia.
<u>8</u>. What is isokinetic sampling?

PROBLEMS AND LABORATORY EXPERIMENTS

<u>13.1</u> Describe in detail all of the equipment necessary to set up and analyze hourly the ambient $SO_2$ concentration for a period of 48 hours.

13.2  Describe in detail all of the equipment
required to determine the $SO_2$ content in a stack gas.
13.3  Determine the isokinetic sampling rates for
a 5-foot diameter duct handling a volumetric gas
flow of 25,000 cfm at 400°F.  The velocity
distribution in the pipe may be considered as 0% of
maximum velocity at the pipe wall, 85% at 0.1
diameters from the wall, 98% at 0.2 diameters from
the wall and 100% at and beyond 0.3 diameters.  Use
the equal area method suggested in Section 13.2 to
locate the sampling points.  The sample probe nozzle
is 3/8 inch I.D.  (Assume no temperature drop.)
13.4  If the sampling system of 13.3 measures the
gas flow at a temperature of 85°F, what flow
reading must be indicated by the measuring device
to assure isokinetic sampling at each point if the
gas density is 2.0 x $10^{-4}$ g/cm$^3$ and the specific
heat is 0.30 Btu/(lb°F)?
13.5  If the dust load in the gas of Problem 13.3 is
1 grain/ft$^3$ at stack conditions and has a mean
diameter of 0.4µ and a standard deviation of 1.9,
how much dust would be collected per minute, a) if
a cyclone separator and a fiber filter were used to
collect the dust and  b) how much dust would be
caught by each of five impingers if the impingers
were designed to capture the size fractions:  less
than 0.4 micron; 0.4-1 micron; 1-4 microns 4-10
microns; and greater than 10 microns?
13.6  Calibrate a 0-8 liter per minute flow meter
using either a wet gas test meter, calibrated
rotometer or calibrated nozzles.
13.7  Determine the volumetric flow in an air tube
using a pitot tube and inclined manometer.  The
equation for determining air velocity near standard
conditions is:

$$\text{air velocity in ft/min} = 1,096.2 \sqrt{\frac{P}{\rho}} \quad (13.8)$$

   where:  P = av. pitot tube pressure differential,
              inches of water
           ρ = air density, lb/ft$^3$

13.8  Determine the mean diameter and size distribution of ammonium chloride (made by mixing vapors
leaving HCl and $NH_3$ bubblers) or cigarette smoke or
any other conveniently generated aerosol using an
inertial impactor.
13.9  Measure the $SO_2$ content of the ambient air
using the West and Gaeke method of analysis.  Check

the results of the $SO_2$ instrumental analyzer with these results.

13.10 Determine the soiling index of the air using a paper tape sampler or a 1-inch diameter filter holder.

13.11 Analyze the exhaust from an automobile for percent carbon monoxide at idle, cruise, acceleration and deceleration and make recommendations for reducing the amount of pollutants emitted.

13.12 Analyze the nitric oxide in a burner flame using the Saltzman wet chemistry technique.

13.13 Absorbents for removing NO, $NO_2$, and $N_2O_4$ include water, alkali solutions, sulfuric acid and phosphoric acid. These oxides can be adsorbed by silica gel, dry sodium ferrate, magnesium dioxide, solid lime and activated carbon. (Activated carbon can adsorb up to 25% of its weight of $NO_2$; it will also adsorb NO and $O_3$.) Pass the gases used in 13.12 through an adsorption or absorption system and measure the amount of nitric oxides in the effluent gas. Compare at least two different removal systems.

13.14 Sulfur dioxide can be removed by catalytic oxidation processes, absorption and adsorption. Absorbents include water, ammonia, and other alkali solutions. Adsorbents include alkalized alumina (a co-precipitate of sodium and aluminum oxides), lignite ash, other mixed metal oxides and activated carbon (activated carbon can adsorb up to 20% of its weight of $SO_2$). Pass a measured quantity of an $SO_2$-air system through an absorption or adsorption system and measure the amount of $SO_2$ in the effluent gas. Compare at least two different removal processes. (Do not use the methods of problem 13.9.)

REFERENCES

1. Byers, R.L., et. al., "A New Technique for Characterizing Atmospheric Aerosols for Size and Shape," Center for Air Environment Studies Publication No. 163-70, The Penn. State Univ., 24 pp. (1970)

2. Smith, W.S., R.T. Shigehara, and W.F. Todd, "A Method to Interpret Stack Sampling Data," 63rd Annual Air Pollution Control Association Meeting, Paper 70-34 (1970)

3. Byers, R.L., and J.W. Davis, "Sulfur Dioxide Adsorption and Desorption on Various Filter Media," JAPCA, Vol. 20 No. 4, pp. 236-238 (1970)

4.  McCrone, W.C., "The Particle Atlas," 1st Edition, Ann Arbor Science Publishers (1967)

5.  Kent, J.A. (ed.), "Riegel's Industrial Chemistry," Reinhold Publishing Corporation (1966)

6.  Munro, L.A., "Chemistry in Engineering," Prentice Hall (1964)

7.  "Selected Methods for the Measurements of Air Pollutants," U.S. Department of Health, Education, and Welfare, Public Health Service Publication No. 999-AP-11 (1965)

8.  "Atmospheric Emissions from Sulfuric Acid Manufacturing Processes," U.S. Department of Health, Education, and Welfare, Public Service Publication No. 999-AP-13 (1965)

9.  "Atmospheric Emissions from Nitric Acid Manufacturing Processes," U.S. Department of Health, Education, and Welfare, Public Health Service No. 999-AP-27 (1966)

10. Cooper, D.W., and R.L. Byers, "Laser Light Backscattering from Laboratory Aerosols," JAPCA, Vol. 20, No. 1, pp. 43-47 (1970)

# APPENDIX A

# DEFINITIONS OF TERMS USED IN AIR POLLUTION CONTROL

ABSORPTION: A process whereby a material extracts one or more substances present in an atmosphere or mixture of gases or liquids; accompanied by physical change, chemical change, or both, of the material.

ACIDIC SMUTS: Acid smuts are solid and liquid conglomerates formed by the condensation of water vapor and sulfur trioxide on cold metal surfaces. They are frequently caused by combustion flue gases coming in contact with a surface whose temperature is below the dew point of the flue gas. They contain metallic sulfate and carbonaceous particles and are approximately one-quarter inch (6,350 microns) in size.

ADIABATIC LAPSE RATE: Adiabatic temperature drop with height of a rising (or falling) parcel of air (-5.4°F/1000 ft.)

ADSORPTION: A physical process in which the molecules of either a gas or a liquid are condensed on the surface of a solid material such as activated carbon, silica gel, etc. Commercial adsorbent materials have enormous internal surfaces.

AEROSOLS: Finely divided solid or liquid particles suspended in a gas. Usually considered to range from 50µ down to sub-micron. This includes: dust, mist, fume, fog, haze and smoke.

AFTERBURNER: A device, including an auxiliary fuel burner and a combustion chamber, in which the combustible air contaminants are incinerated.
*Direct Flame Afterburner:* That type in which the auxiliary burner provides all the necessary heat and flame contact necessary for incineration.
*Catalytic Afterburner:* That type in which the surface action of catalysts are employed

such that incineration occurs at a lower
temperature than it would in a direct flame
unit.  Less auxiliary heat is required with
a catalytic afterburner than with a direct
flame afterburner.

*Recuperative Afterburner:*  A direct flame unit
in which a heat exchanger is utilized to
pre-heat the incoming contaminated gases so
that less auxiliary fuel is required.

AIR ATMOSPHERE:  A mixture of invisible, odorless,
tasteless gases surrounding our earth, containing
by volume when dry:  78.09% nitrogen, 20.94%
oxygen, 0.93% argon, 0.03% carbon dioxide, and
traces of the elements neon, helium krypton,
hydrogen and xenon, and the compounds methane
and nitrous oxide.  Moist air atmosphere contains
up to three percent water vapor by volume.
Approximately half of the earth's total air
atmosphere, by weight, lies below the 18,000
foot altitude.

AIR CONTAMINANTS:  Aerosols or gases which are
discharged into the outdoor atmosphere, where
they might exhibit an adverse effect.

AIR MONITORING:  The continuous sampling for and
measuring of the quantity of air pollutants
present in the outdoor atmosphere.

AIR POLLUTION:  Presence of *abnormal* quantities of
certain matter in the atmosphere that are
det mental to the health and/or welfare of man.

AIR QUALITY STANDARDS:  A level beyond which air
pollutants in the atmosphere can cause injury
to plants, animals or materials.  Both concentration and time factors are considered in
establishing standards.  Standards are determined
using "criteria" published by OAP EPA.

AIR RESOURCE MANAGEMENT:  The establishment and
enforcement of specified objectives supported
by appropriate planning, control regulations,
laboratory and testing facilities, and effective
enforcement agencies and procedures to prevent
and control the contamination by pollution of
our outdoor air environment.

AMBIENT AIR:  The surrounding local air.

APCO:  Air Pollution Control Office, the first
name used for OAP EPA (formerly NAPCA).

ATOM:  The smallest particle of an element that can
exist either alone or in combination with other
atoms of the same or of another element.

AUXILIARY FUEL FIRING EQUIPMENT:  Equipment to
supply additional heat to incinerators by the

combustion of an auxiliary fuel for the purpose
of obtaining temperatures sufficiently high:
(a) to dry and ignite the waste material, (b)
to maintain ignition thereof, and (c) to promote
complete combustion of combustible solids,
vapors and gases.

CAPACITY, TOTAL LUNG (TLC): The volume of gas
contained in the lungs at full inspiration.

CAPACITY, VITAL: The maximum volume of gas which
can be expired from the lungs following maximum
inspiration.

CARCINOGEN: A substance capable of causing living
tissue to become cancerous.

CHLOROSIS: The loss of chlorophyll in a plant which
produces sickening and yellowing of vegetation.

CHUTE, CHARGING (INCINERATOR): A pipe or duct
through which wastes are conveyed by gravity
from above to a primary chamber or to storage
facilities preparatory to burning.

COH/1000 LINEAR FEET (COEFFICIENT OF HAZE PER 1000
LINEAR FEET): A measure of the optical density
of atmospheric dust or other particulate matter
as deposited on a filter paper. ASTM Standard
D 1704-61 gives:

$$COH/1000 \text{ linear feet} = \frac{(ft.^2 \text{ tape area})(100,000)}{(ft.^3 \text{ air sampled})} \log \frac{100}{\% \text{ transmission}}$$

CLOTH FILTER COLLECTORS: A mechanical filtration
system for removing solid pollutants from a
gas stream. They operate on a filtration
principle similar to that of the ordinary
vacuum cleaner bag. They are often called bag-
type collectors since their most common
arrangement is in a form of a number of
cylindrical bags mounted vertically within an
enclosure. Cloth collectors are capable of
removing sub-micron size particles with
cleaning efficiencies in excess of 99%.

COALESCENCE: In cloud physics, the merging of two
drops of liquid into a single large drop.

COARSE SOLID PARTICLES: Solid particles having a
size equal to or greater than 50 microns, and
solid particles contained in or on liquid
particles.

COMBUSTION: The chemical combination of oxygen
and combustible matter resulting in the rapid
release of energy and products of combustion

(incompletely burned organic species, CO, $CO_2$, $H_2O$, $NO_x$, etc.).

CRITERIA:  Information used as guidelines for establishing air quality standards or goals.

CYCLONE COLLECTORS:  A mechanical system employing centrifugal force for removal of aerosols (liquids or wet or dry solids) from gas streams by imposing a rapid whirling motion to the entering gas stream.

CONIFER:  An evergreen type tree or shrub.

DISPERSION:  The most general term for a system consisting or particulate matter suspended in air or other gases.

DISPERSOID:  The particles in a dispersion.

DISTILLATE OILS:  Oils which have been separated from crude oils by fractional distillation. Generally contain less sulfur and are less viscous than residual oils.

DUST:  Finely divided particles which are formed by mechanical disintegration (size = one to several hundred microns). Dust particles usually are irregular in shape.

ECOLOGY:  The relation and interacting of living organisms with their surroundings.

EFFECTIVE STACK HEIGHT:  The sum of the stack height and the plume rise.

ELECTROSTATIC PRECIPITATOR:  An electrical system for the removal of aerosols from a gas stream by giving them an electrical charge, then collecting them at a plate imparted with an opposite charge.

EMISSION:  The material which is being discharged into the outdoor atmosphere.

EMISSION FACTOR:  A typical value to indicate normal amounts of pollutants released from a given emission source when operated using specified control procedures and devices.

EMISSION STANDARDS:  The amount of pollutants that are permitted to be discharged from an air pollution source.  The following are the more popular ways to describe the quantity of air pollutants being discharged.

a.  Weight of pollutants per unit time.

b.  Pollutant weight per unit volume or weight of effluent gas.

c.  Weight of pollutants per unit weight of material processed.

d.  Volume of pollutants per unit volume of effluent gas.

EMPHYSEMA, PULMONARY: An overdistension of lung
air spaces resulting in destruction of the lung
alveoli and other functioning tissues.
ENVIRONMENTAL SCIENCE: The study of life and its
surroundings and their interrelationships
directed toward the improvement of man's well
being.
EPA: Environmental Protection Agency. (See Appendix
B).
EPIDERMIS: Outer covering (skin) on animals or
plants.
EXCESS AIR: That air supplied in addition to the
theoretical quantity required for complete
combustion of all the fuel and/or waste present.
FLARE: A device utilized at petroleum refineries,
heat treating operations, liqu fied petroleum
gas facilities, etc., to burn rich mixtures of
combustible waste gases. A flare differs from
"afterburner" in that no auxiliary fuel other
than a pilot flame is necessary to incinerate
the gases.
FLUE: A passage for conducting products of combustion into the atmosphere.
FLUE-FED INCINERATOR: An incinerator which is
charged (refuse is conveyed by gravity to the
primary chamber) through a vertical flue which
also serves as the passage for conducting
products of combustion to the atmosphere.
FLY ASH: The finely divided particles of ash
entrained in flue gases arising from the combustion of fossil fuels (predominantly coal). The
particles of ash may contain incompletely
burned fuel.
FOSSIL FUELS: Coal, petroleum and natural gas.
FUMES: Solid particles generated by condensation
of metals from a gaseous state. (Size 0.001
to 1 microns). Usually spherical in shape.
HEW: U.S. Department of Health Education and
Welfare.
HYDROCARBONS: Compounds containing only hydrogen
and carbon.
IMPACTOR, CASCADE: An instrument consisting of
stages for producing successively increasing
air velocities for collecting particles by
size ranges.
INSOLATION: A rate at which direct solar radiation
reaches the earth's surface.
INVERSION: A condition occurring when the lapse rate
is greater than the adiabatic lapse rate (-5.4°F/
1000 ft.). Under these conditions the stagnant
air pollution concentrations begin to build up.

INCINERATION: The process of burning solid, semi-
solid or gaseous combustible wastes to an
inoffensive gas and a sterile residue containing
little or no combustible material.

MEAN, GEOMETRIC: The nth root of the product of n
terms.

MICRON: 1 micron ($\mu$) is $10^{-6}$ meters ($10^{-4}$ cm.).

NAPCA: National Air Pollution Control Administration,
formerly a division of HEW and now joined with
EPA (see Appendix B). For publications write
either: Technical Publications Division, 1033
Wade Ave., Raleigh, N.C. 27605; or U.S. Depart-
ment of Commerce, National Bureau of Standards,
Clearinghouse for Federal Scientific and
Technical Information, Springfield, Va. 22151.

NATURAL ATMOSPHERIC DISPERSOIDS, SIZE OF: Although
there is no distinct size line between categories,
the approximate particle size range as given
in the "Glossary of Meteorology" of the American
Meteorological Society is as follows:
Haze:  1 $\mu$
Mist:  1-5 $\mu$
Cloud or Fog:  5-200 $\mu$
Drizzle:  200-500 $\mu$
Rain:  500-8,000 $\mu$
Also, see "Aerosol".

NECROSIS: Collapse and death of tissue

OAP: Office of Air Programs of EPA (See Appendix B).

ODORS: Substances that stimulate the olfactory
organ, causing the sensation we call "smell".
*Malodors:* An odor which is considered unpleasant
by most humans.

OPACITY: The degree to which a plume or column or
exhaust gases obscures an observer's view.
Opacities are reported in terms of percent
obscuration such that a 100% opacity plume is
one which completely obscures one's line of
sight through the plume.

OPEN BURNING: Any fire wherein the products of
combustion are emitted into the open air, and
are not directed thereto through a stack or
chimney.

OXIDANTS: Those substances in the air (such as
ozone, PAN, and nitrogen dioxide) which are
capable of oxidizing other chemicals or elements
in oxidation-reduction type chemical reactions.
Oxidants are made up mostly of ozone, and only
small amounts of the other materials.

PALISADE TISSUE: The layer of columnar cells directly
beneath the leaf upper epidermis. They are rich
in chloroplasts.

PAN: (Peroxyacetyl nitrate--$CH_3COONO_2$ or peroxyacyl nitrate.) Produced in the atmosphere by photochemical reaction between the oxides of nitrogen and hydrocarbons under the influence of sunlight. PAN is an important phytotoxicant.

PARTICULATE MATTER: Any liquid or solid matter in the atmosphere. Size may range from $0.0002\mu$ to $500\mu$ in diameter.

PARTS PER MILLION (PPM): Number of parts of gas, by volume, in a million total parts. (Volume ratio = mole ratio = pressure ratio for ideal gases.).

PHOTOCHEMICAL REACTIONS: Chemical reactions occuring in the atmosphere under the influence of sunlight. The energy for these reactions is obtained from sunlight, particularly the ultraviolet rays. The results of these reactions are photochemical smogs.

PHYTOTOXIC: Poisonous to plants.

PLUME: The flow of visible effluent from a specific outlet, such as a stack or vent.

RESIDUAL OILS: Oils which are too heavy to be evaporated in any normal evaporation or distillation process and are left over from that process.

RINGELMANN'S CHART: It consists of four, five and three-quarter by eight and one-half inch charts, each with a rectangular grid of black lines on a white background. The width and spacing of the lines are designed so that each chart presents a certain percentage of black. Ringelmann #1 is equivalent to 20% black; Ringelmann #2, 40% black; Ringelmann #3, 60% black; Ringelmann #4, 80% black; Ringelmann #5, which is not part of the chart, is 100% black. It is used to evaluate the degree of blackness of smoke plumes.

SC: See "Standard Conditions".

SCRUBBERS AND ABSORBERS: Wet systems that are used for removing aerosol and gaseous pollutants from an air stream. The gases are removed by absorption and perhaps reaction. The solid and liquid particles are removed through wet contact. High energy scrubbers are capable of cleaning efficiencies of 99% or more for particulates ranging in size from sub-micron and larger.

SMOG: A term derived from a combination of the words "smoke" and "fog". It is used to characterize certain types of atmospheric

pollution by aerosols. However, this term has
not been given a precise definition.

*London Smog:*   Condition occuring in cooler,
  foggy weather during periods or air stagnation
  in areas where coal is the principle space
  heating fuel.  It is characterized by the
  high content of sulfur compounds, smoke and
  fly ash, and may produce extremely poor
  visibility, bronchial irritation, and has
  been known to cause death.

*Photochemical Smog:*   A condition occuring in
  sunny, poorly ventilated, heavily-motorized
  urban areas.  It is characterized by the
  interaction of oxides of nitrogen and
  hydrocarbons (one of the major sources of
  these is the automobile) under the influence
  of sunlight.  May produce poor visibility,
  eye irritation, and damage to materials and
  vegetation.

SMOKE:  Small gas borne liquid and/or solid particles
  formed by incomplete combustion, consisting
  predominantly of carbon and other combustible
  material, such as ash.  (Size = 0.001 to 1.0
  microns).

SOOT:  A conglomeration of particles impregnated
  with "tar", formed by the incomplete combustion
  of carbonaceous material.

SOURCE SAMPLING:  Testing of the quality and
  quantity of an emission directly from an air
  pollutant generating source.  The collection
  is made within the gas stream emitting from
  the source before any ambient air dilution
  has taken place.

STAGNATING ANTICYCLONES:  Large, high pressure
  air masses which tend to remain over an area for
  a period of four days or longer.  Clear skies,
  low wind velocities, abnormally warm weather,
  cool nights, and high air pollution potential
  are normally associated with these air masses.
  In the Mid-Atlantic States area they are most
  prevalent in the fall.  The weather associated
  with the fall stagnating anticyclones is
  commonly termed "Indian Summer" weather.

STANDARD:  The maximum level of air contaminant as
  established by a legal authority.

STANDARD CONDITIONS (SC):  Conditions of 70°F and
  one atmosphere (14.7 psia or 760 mm Hg), unless
  otherwise noted in this publication

STOICHIOMETRIC:  The exact quantity of reactants
  required to react according to a particular

chemical equation. If the reaction were complete, only products and no reactants would remain.

STP: Standard temperature and pressure of 32°F and 1 atmosphere. Values such as scfh (standard cubic feet per hour) are based on STP.

STOKES' DIAMETER: The equivalent spherical diameter of the particle being considered.

STOKES' LAW: The kinetic force required to overcome form and friction drag of a sphere moving through a fluid is proportional to the fluid viscosity, sphere radius and relative velocity of sphere and fluid ($F_k = 6\pi\mu rv$).

TAR: A thick dark brown or black viscous liquid obtained by distillation of wood, coal, peat, etc.

TERMINAL SETTLING VELOCITY: Derived using Stokes' Law and Newton's drag coefficient relations at Reynolds No. <0.1. It shows that the terminal free fall velocity of a sphere moving through a fluid is proportional to the sphere radius squared and density, and inversely proportional to the fluid viscosity.

THEORETICAL AIR: The chemically correct quantity of air required for complete combustion of a given quantity of a specific fuel without excess air. Also designated as Stoichiometric Air.

TIDAL VOLUME: Amount of gases inhaled or exhaled each breath (maximal breathing capacity ≅ 125 ℓ gases breathed in each minute).

TLV (THRESHOLD LIMIT VALUE OR MAC--MAXIMUM ALLOWABLE CONCENTRATION): Refers to indoor airborne concentrations of substances and represents conditions under which it is believed that nearly all workers may be exposed eight hours a day, six days per week without adverse effects. A list of recommended and intended values is compiled by the American Conference of Governmental Industrial Hygienists.

VAPORS: The gaseous phase of matter which normally exists closer to the equilibrium vaporization line and therefore is *less* like "ideal gases" than are gases.

VENTILATION: The movement and circulation of outdoor air commonly termed "wind". These vertical and horizontal air currents over an area can act to carry away and disperse air pollutants and thereby prevent a build-up of the pollutants in the general area of their production.

# APPENDIX B

## SOURCES FOR AIR POLLUTION ASSISTANCE

It is frequently necessary to obtain technically qualified assistance in solving air pollution problems. If this is necessary, the following suggested list may be of benefit:

(1)   American Academy of Environmental Engineers, Environmental Engineering Inter-Society Board, Inc.,
      P. O. Box 9728
      Washington, D.C.   20016

      This organization is composed of registered professional engineers who have also passed the society's environmental specialty tests.

(2)   Air Pollution Control Association (APCA)
      4400 Fifth Avenue
      Pittsburgh, Pa.   19213

      This association is a national organization devoted solely to air pollution control and currently has over 5,000 members who are active in all aspects of air pollution control. A directory of members may be obtained from the association headquarters.

(3)   APCA Consultant Guide--This directory of air pollution control consultants lists names and addresses according to five broad geographic areas in the United States. This guide of consultants can be obtained from the association headquarters.

(4) Directory of Governmental Air Pollution Agencies-
This directory is published by the Air Pollution
Control Association in cooperation with the
United States Department of Health, Education
and Welfare, and lists both federal and state
agencies for the United States and Canada.

(5) National Society of Professional Engineers
(NSPE)
2029 K Street, N.W.
Washington, D.C.   20006

NSPE publishes the "Directory of Engineers in
Private Practice" (publication number NSPE No.
1919) which includes listings of professional
engineers in private practice who are specialists
in air pollution as well as other related areas
such as solid waste disposal, water treatment.

(6) State Professional Engineering Societies--
Rosters listing the names of professional
engineers can be obtained for a particular
state from the state capital. The only state
that specifically registers environmental
engineers is the state of Pennsylvania, however,
it is expected that other states will follow
suit at a later date.

(7) Chemical Engineering--This McGraw-Hill publication publishes an annual environmental
engineering issue which includes, among other
things, a summary of the current federal and
state pollution control laws.

The above list is by no means conclusive, however,
it may serve as a starting point for locating
qualified persons to assist in air pollution control
problems.
Technical information or literature in air
pollution available through the Federal Government
may be obtained from:

(1) Environmental Protection Agency (EPA)
Office of Air Programs (OAP)--formerly National
Air Pollution Control Administration (NAPCA)
1033 Wade Avenue
Research Triangle
Raleigh, N.C.,   27605

(2)   EPA, OAP
      Office of Education and Information,
      Park Lawn, 5600 Fishers Lane,
      Rockville, Md.   20852

(3)   Superintendent of Documents,
      Government Printing Office
      Washington, D.C.   20402

(4)   U.S. Department of Commerce,
      National Bureau of Standards,
      Clearinghouse for Federal Scientific &
       Technical Information,
      Springfield, Va.   22151

   Information may also be obtained from state
and local air pollution control agencies, and the
local chapters of the National Tuberculosis and
Respiratory Disease Association, and the American
Cancer Society.
   A suggestive listing of where air pollution
control equipment, analytical instruments,
accessories, and chemicals may be obtained is
included in Section 13.4.

# APPENDIX C

## CONVERSION FACTORS

LENGTH

1 inch = 2.54 cm
1 mile = 5280 ft = 1609 m
1 micron = $10^{-3}$mm = $10^{-6}$m
$1A^\circ$ = $10^{-8}$cm = $10^{-10}$m
1m = 3.28 ft
1 ft = 0.305 m

MASS

1 lb = 453.6 g
1 ton = 2000 lb
1 lb = 7000 grains = 16 oz

GEOMETRY

circle area = $\pi r^2$ = $\pi d^2/4$
cylinder volume = $\pi r^2$ x height
sphere area = $4 \pi r^2$ = $\pi d^2$
sphere volume = $4/3 \pi r^3$ = $1/6 \pi d^3$

VISCOSITY

1 poise = 1 g/(cm sec) = absolute viscosity
1 centipoise = 0.000672 lb/(sec ft)
1 stoke = $cm^2$/sec = poise/$\rho$ = kinematic viscosity

ACCELERATION

g = 32.174 ft/$sec^2$ = 980.665 cm/$sec^2$ at sea level

## PRESSURE

$$1 \text{ atm} = 1.01325 \text{ bar} = 14.696 \text{ lb}_f/\text{in}^2$$
$$= 29.92 \text{ in Hg } (32°F) = 760 \text{ mm Hg } (0°C)$$
$$= 33.936 \text{ ft H}_2O \ (60°F) = 760 \text{ torr}$$

## DENSITY OF:

air $= 1.2 \times 10^{-3} \text{g/cm}^3 = 7.43 \times 10^{-2} \text{lb/ft}^3$
     at Standard Conditions
water $= 62.4 \text{ lb/ft}^3 = 1 \text{ g/cm}^3$ at 4°C
mercury $= 13.6 \text{ g/cm}^3$ at 4°C
glass $= 2.2 \text{ g/cm}^3$
cement $= 1.5 \text{ g/cm}^3$
earth $\cong 1.5 \text{ g/cm}^3$
limestone $= 2.1\text{-}2.9 \text{ g/cm}^3$
sandstone $= 2.0\text{-}2.6 \text{ g/cm}^3$
coal $= 1.0\text{-}1.5 \text{ g/cm}^3$
iron ore $= 3.5\text{-}5.0 \text{ g/cm}^3$
iron slag $= 2.5\text{-}3.0 \text{ g/cm}^3$

## TIME

1 hr = 3600 sec
1 work year $\cong$ 2000 hours per person
1 year $\cong$ 8000 total hours

## VOLUME

1 ft$^3$ = 7.481 U.S. gal = 28.32 $\ell$
1 gal = 8.35 lb $H_2O$ @ S C
1 barrel = 42 gal oil
1 liter = 1.057 quarts
1 liter = 0.2642 gal
1 cm$^3$ = 1 ml
1 m$^3$ = 35.314 ft$^3$ = 1000 $\ell$

## FORCE

1 dyne = g cm/sec$^2$
1 lb$_f$ = lb$_m$ g/g$_c$

## ENERGY

1 erg = dyne cm
1 joule = $10^7$ erg
1 cal = 4.184 joule
1 cal/g mole = 1.8 Btu/lb mole
1 Ton refrigeration = 200 Btu/min
1 Btu = 252.2 cal = 778.2 ft lb$_f$

1 kw hr = 3412 Btu = 1.341 hp hr
1 hp hr = $1.98 \times 10^6$ ft $lb_f$

POWER

1 watt = 1 joule/sec = $10^7$ erg/sec
1 hp (mech) $\cong$ 746 watt = 33,000 ft $lb_f$/min = 2,545
   Btu/hr
1 hp (boiler) = 13.155 hp (mech)
1 hp (boiler) $\cong$ 34.5 lb steam/hr $\cong$ 2.5 lb oil/hr
   (at 75% thermal efficiency)

TEMPERATURE

$^\circ$F = 1.8$^\circ$C + 32 = 2$^\circ$C-0.2$^\circ$C + 32
$^\circ$R = $^\circ$F + 459.49 = 1.8$^\circ$K
$^\circ$K = $^\circ$C + 273.16

SPEED

1 ft/sec = 30.48 cm/sec = 0.6818 mi/hr
1 mi/hr = 0.447 m/sec

CONSTANTS

$g_c$ = 32.174 $lb_m$ ft/ ($lb_f$ sec$^2$) = 980.7 cm/sec
e = 2.7183
$\pi$ = 3.14159
ln 10 = 2.3026
Boltzman's constant = $K$ = $1.38 \times 10^{-16}$ g cm$^2$/(sec$^2$
   molecule $^\circ$K) = $1.38 \times 10^{-23}$ joules/$^\circ$K
Avogadro's Number = $6.02 \times 10^{23}$ atoms/g atom
Gas constants = R = 1.987 cal/(g mole $^\circ$K) = 82.05
   atm cm$^3$/(g mole $^\circ$K) = $4.968 \times 10^4$ $lb_m$ft$^2$/(lb
   mole $^\circ$R)
1 lb mole = 359 ft$^3$ ideal gas at STP
1 g mole = 22.4 liter ideal gas at STP
$(C_p)_{H_2O}$ $\cong$ 1 Btu/$lb_m$$^\circ$R at S C
$(C_p)steel$ $\cong$ 0.2 Btu/$lb_m$$^\circ$R at S C
1 coulomb = $6.2 \times 10^{18}$ electrons = $1.038 \times 10^{-4}$
   mole Ag+ ion = 1 amp sec
1 amp = 1 coulomb/sec
1 esu = $1.59 \times 10^{-19}$ amp sec = $1.59 \times 10^{-19}$ coulombs

AIR AT STANDARD CONDITIONS

$\mu_a$ = $1.83 \times 10^{-4}$g/(cm sec)
   = $3.76 \times 10^{-7}$ $lb_f$ sec/ft$^2$
   = $1.21 \times 10^{-5}$ $lb_m$/(sec ft)

$(C_p)_{air}$ = 0.26 Btu/$(lb_m °R)$ = 0.26 Cal/$(g°K)$
$M_{air}$ = 28.96 $\cong$ 29
$k_a$ = 4.02 x $10^{-6}$ Btu/(sec ft °R)
$\rho_a$ = 1.19 x $10^{-3} g/cm^3$ = 7.50 x $10^{-2}$ $lb_m/ft^3$
Composition $\cong$ 21.0% $O_2$ and 79.0% $N_2$ by vol
$\qquad\qquad\quad \cong$ 23.2% $O_2$ and 76.8% $N_2$ by mass

CONVERSION FACTORS FOR AIR POLLUTION

Convert ppm to wt per vol ratio at SC

general:   (ppm) $\dfrac{M}{0.0241}$ = µg/m

specific:

*mult ppm at SC*      *by listed value to obtain* µg/m³

| | mult ppm at SC → | µg/m³ |
|---|---|---|
| $CH_4$ | | 663 |
| CO | | 1160 |
| $CO_2$ | | 1820 |
| HCN | | 1120 |
| HF | | 828 |
| $H_2S$ | | 1410 |
| $NO_2$ | | 1910 |
| $O_3$ | | 1990 |
| $SO_2$ | | 2650 |

1 acre = 43,560 $ft^2$
1 number/$ft^3$ = 35.31 number/$m^3$
1 ton/$mi^2$ = 3.125 lb/zcre = 0.0717 $lb/ft^2$
1 $g/m^3$ = 0.0283 $g/ft^3$
1 lb/hr = 0.126 g/sec

# SUBJECT INDEX

Printed and bound by
LithoCrafters, Inc.
Ann Arbor, Michigan
USA